"Enjoyable . . . This is a good entry point for newcomers to this long-running series." —*Publishers Weekly*

Class Act

"Fast-paced . . . The rising suspense keeps readers riveted, as does the body count." —Bookreporter.com

Double Jeopardy

"A clever and captivating reading experience."

—Bookreporter.com

Hush-Hush

"Prime escapism from a master of the genre." —*Booklist*

Shakeup

"Woods's alternate universe . . . provides perfect escapist fare." —*Publishers Weekly*

Choppy Water

"This is prime Woods . . . . Slick, reliable, action-oriented escapism, just right for troubled times." —*Booklist*

Hit List

"Fast-moving easy reading in the familiar Woods style."

—*Booklist*

Treason

"Bon appétit to readers who like their international intrigue sweet and weightless as a soufflé." —*Kirkus Reviews*

Stealth

"Smooth, easy-reading escapism in the trademark Woods style." —*Booklist*

Contraband

"Woods balances the criminal doings with loads of light banter and tantalizing glimpses of great sex."

—*Publishers Weekly*

Skin Game

"The ingenious ways in which Teddy outwits his adversaries is a large part of this entry's appeal. Series fans will hope he has a long career." —*Publishers Weekly*

Wild Card

"Woods shows no sign of flagging in this long-running series." —*Publishers Weekly*

A Delicate Touch

"Woods continues to deliver satisfying escapist fare, with the Bondian Barrington outmaneuvering his foes yet again, always with high-tech gadgetry at the ready and a beautiful woman at his side." —*Booklist*

Desperate Measures

"Woods creates another action-packed thriller for his readers to devour, with plenty of interesting twists and turns that make for a nonstop, can't-catch-your-breath read." —*Booklist*

Turbulence

"Inventive . . . [An] alluring world of wealth, power, and crime." —*Publishers Weekly*

Shoot First

"Smooth . . . The principal pleasure lies in watching the suave, resourceful Stone maintain his good humor and high lifestyle throughout all his travails." —*Publishers Weekly*

Unbound

"Stuart Woods is a no-nonsense, slam-bang storyteller." —*Chicago Tribune*

Quick & Dirty

"Suspenseful . . . The excitement builds."

—*Publishers Weekly*

Indecent Exposure

"[An] irresistible, luxury-soaked soap opera."

—*Publishers Weekly*

Fast & Loose

"Another entertaining episode in [a] soap opera about the rich and famous." —Associated Press

Below the Belt

"Compulsively readable . . . [An] easy-reading page-turner."

—*Booklist*

Sex, Lies & Serious Money

"Series fans will continue to enjoy this bird's-eye view of the high life." —*Booklist*

Dishonorable Intentions

"Diverting." —*Publishers Weekly*

Family Jewels

"Mr. Woods knows how to portray the beautiful people, their manners and mores, their fluid and sparkling conversation, their easy expectations, and all the glitter that surrounds and defines them. A master of dialogue, action, and atmosphere, [Woods] has added one more jewel of a thriller-mystery to his ever-growing collection."

—*Florida Weekly* (Fort Myers)

Scandalous Behavior

"Addictive . . . Pick [*Scandalous Behavior*] up at your peril. You can get hooked." —*Lincoln Journal Star*

Foreign Affairs

"Purrs like a well-tuned dream machine. . . . Mr. Woods knows how to set up scenes and link them to keep the action, emotion, and information moving. He presents the places he takes us to vividly and convincingly. . . . Enjoy this slick thriller by a thoroughly satisfying professional."

—*Florida Weekly*

Hot Pursuit

"Fans will enjoy the vicarious luxury ride as usual."

—*Publishers Weekly*

Insatiable Appetites

"Multiple exciting storylines . . . Readers of the series will enjoy the return of the dangerous Dolce." —*Booklist*

Paris Match

"Plenty of fast-paced action and deluxe experiences that keep the pages turning. Woods is masterful with his use of dialogue and creates natural and vivid scenes for his readers to enjoy." —*The Sun News*

Cut and Thrust

"This installment goes down as smoothly as a glass of Knob Creek." —*Publishers Weekly*

BOOKS BY STUART WOODS

FICTION

*Distant Thunder**
*Black Dog**
*A Safe House**
*Criminal Mischief**
*Foul Play**
*Class Act**
*Double Jeopardy**
*Hush-Hush**
*Shakeup**
*Choppy Water**
*Hit List**
*Treason**
*Stealth**
*Contraband**
*Wild Card**
*A Delicate Touch**
*Desperate Measures**
*Turbulence**
*Shoot First**
*Unbound**
*Quick & Dirty**
*Indecent Exposure**
*Fast & Loose**
*Below the Belt**
*Sex, Lies & Serious Money**
*Dishonorable Intentions**
*Family Jewels**
*Scandalous Behavior**
*Foreign Affairs**
*Naked Greed**
*Hot Pursuit**
*Insatiable Appetites**
*Paris Match**
*Cut and Thrust**
*Carnal Curiosity**
*Standup Guy**
*Doing Hard Time**
*Unintended Consequences**
*Collateral Damage**
*Severe Clear**
*Unnatural Acts**
*D.C. Dead**
*Son of Stone**
*Bel-Air Dead**
*Strategic Moves**
Santa Fe Edge†
*Lucid Intervals**
*Kisser**
Hothouse Orchid‡
*Loitering with Intent**
Mounting Fears§
*Hot Mahogany**
Santa Fe Dead†
Beverly Hills Dead
*Shoot Him If He Runs**
*Fresh Disasters**
Short Straw†
*Dark Harbor**
Iron Orchid‡
*Two Dollar Bill**
The Prince of Beverly Hills
*Reckless Abandon**
Capital Crimes§
*Dirty Work**
Blood Orchid‡
*The Short Forever**

Orchid Blues‡

*Cold Paradise**

*L.A. Dead**

The Run§

*Worst Fears Realized**

Orchid Beach‡

*Swimming to Catalina**

*Dead in the Water**

*Dirt**

Choke

Imperfect Strangers

Heat

Dead Eyes

L.A. Times

Santa Fe Rules†

*New York Dead**

Palindrome

Grass Roots§

White Cargo

Deep Lie§

Under the Lake

Run Before the Wind§

Chiefs§

COAUTHORED BOOKS

*Obsession*** (with Brett Battles)

*Near Miss** (with Brett Battles)

*Jackpot***
(with Bryon Quertermous)

*Bombshell*** (with Parnell Hall)

*Skin Game*** (with Parnell Hall)

*The Money Shot***
(with Parnell Hall)

Barely Legal†† (with Parnell Hall)

*Smooth Operator***
(with Parnell Hall)

AUTOBIOGRAPHY

An Extravagant Life

TRAVEL

A Romantic's Guide to the Country Inns of Britain and Ireland (1979)

MEMOIR

Blue Water, Green Skipper

**A Stone Barrington Novel*

†*An Ed Eagle Novel*

‡*A Holly Barker Novel*

§*A Will Lee Novel*

***A Teddy Fay Novel*

††*A Herbie Fisher Novel*

NEAR MISS

STUART WOODS

and BRETT BATTLES

G. P. PUTNAM'S SONS

NEW YORK

PUTNAM
— EST. 1838 —

G. P. PUTNAM'S SONS
Publishers Since 1838
An imprint of Penguin Random House LLC
penguinrandomhouse.com

The Library of Congress has catalogued the G. P. Putnam's Sons hardcover edition as follows:

Names: Woods, Stuart, author. | Battles, Brett, author.
Title: Near miss / Stuart Woods and Brett Battles.
Description: New York: G. P. Putnam's Sons, [2023] |
Series: A Stone Barrington novel; 64
Identifiers: LCCN 2023012488 (print) | LCCN 2023012489 (ebook) |
ISBN 9780593540060 (hardcover) | ISBN 9780593540077 (ebook)
Subjects: LCSH: Barrington, Stone (Fictitious character)—Fiction. |
Private investigators—Fiction. | LCGFT: Thrillers (Fiction) |
Detective and mystery fiction. | Novels.
Classification: LCC PS3573.O642 N43 2023 (print) |
LCC PS3573.O642 (ebook) | DDC 813/.54—dc23/eng/20230317
LC record available at https://lccn.loc.gov/2023012488
LC ebook record available at https://lccn.loc.gov/2023012489

First G. P. Putnam's Sons hardcover edition / June 2023
First G. P. Putnam's Sons premium edition / February 2024
G. P. Putnam's Sons premium edition ISBN: 9780593540084

Printed in the United States of America
1 3 5 7 9 10 8 6 4 2

NEAR MISS

CHAPTER 1

Stone Barrington stood in the shooting range in the basement of his New York City home, held the Colt Model 1911 semiautomatic .45-caliber pistol out in front of him with both hands, and squeezed off a round.

The president of the United States, Holly Barker, stood next to him. "Near miss," she said.

"I hit him in the chest," Stone protested. "All right, sort of in the chest."

"You should be able to hit him in the center of the chest or in either eye," Holly said. "I don't know why I can't fix you. If my father, Ham, were here, he'd fix you in a minute. Why can't I do that?"

"It just may have something to do with the quality of your student," Stone said.

"You can shoot the eye out of a gnat with a rifle. Why can't I get you to do that with a .45?"

Stone tried again. "Maybe I'm unfixable. Maybe I've been hit over the head too many times, and my brain is wobbly."

"I could always fix anybody, like Ham does."

"Maybe serving as president has cut into your range time," Stone suggested.

"What range time? The Secret Service gets all wobbly anytime I'm in the same room with an unholstered handgun," she said. "I guess they're afraid I'll shoot myself."

"Then they don't understand whose daughter you are."

"Right about that. They're all so . . . *young*. They can't remember as far back as when Ham was national pistol champion two years in a row."

"Maybe Ham has faded a little?"

"Forget about that. He only competed twice. He could win it again today."

"If I ever need somebody shot, I'll get Ham to do it for me," Stone said.

"Good idea."

"I'm fatally limited to having no more than one good idea a day, without a healing infusion of Knob Creek," Stone said. "Let's adjourn to my study."

"Okay," Holly said. "Then, after I've loosened you up, I'll take you upstairs and fuck your brains out, if we can get past the Secret Service."

"You must be the only female president to talk that way," Stone said.

"Wrong. Kate Lee served two terms and never stopped talking that way. She still does."

They let themselves out of the shooting range, past a nervous-looking Secret Service agent, and sought solitude in Stone's study. Stone poured.

They took a swig of the bourbon, then Holly sat back on the sofa and gazed at him. "Stone," she said, "there's something I've never been able to figure out about you."

"What's that? I'm an open book."

"Why are bad people always trying to kill you?"

"I've often wondered about that myself," Stone said. "And I can't figure it out, either."

"Why are you such an inviting target?" she asked.

"Your guess is as good as mine," he replied.

"Dino says it's because you're sloppy. He can't get you to carry all the time, and you're constantly forgetting to set your very expensive alarm system. I mean, Mike Freeman or Bob Cantor is always updating it with the latest feature, yet people keep getting into your house."

"Well, not when you're here," Stone said. "Those Secret Service people are the best alarm system on the planet."

"That's what they think, too. A tip: don't bring up Jack Kennedy or even George Wallace in their presence. They get all defensive."

"Then I'll never bring up those names," Stone promised. He rose and went to get a fresh bottle of Knob Creek from the cabinet next to the bar.

"They'll never need to watch your ass," Holly said. "I'll do that for them."

"That's the nicest thing anyone has said to me in this millennium," Stone said, putting that part of himself back on the sofa.

"Have you loosened up yet?" Holly asked.

"I believe I have," Stone replied.

So, they went upstairs, past two Secret Service agents, and did what Holly had suggested earlier. They did it well, too, being accustomed to each other's bodies.

CHAPTER 2

Stone got Holly into his Bentley, which was nearly as well armored as the car the government supplied for her, and shipped her off to the East Side Heliport. She was wearing a burka that Ralph Lauren's people had made for her, which saved her from recognition by the public. Her helicopter was something smaller than Marine One, and that helped, too. It was normally used by the CIA, but they were happy to loan it to her, especially at budget time.

Stone went back into his study and called Dino Bacchetti.

"Yo."

"'Yo'? Is that the person to whom I am speaking?"

"Yeah, who's this, Lily Tomlin?"

"Dinner at Clarke's, in an hour?"

"You're on." Dino hung up.

They got through a drink at the bar before they were called to their table. There was a very attractive woman seated next to Stone, opposite a handsome young man.

"I need to powder my nose," she said to her companion, and he let her out from behind their table, though not too graciously.

As soon as she had gone, the young man threw some money on the table and got a waiter's attention. "Bring her another drink and tell her I was called away on important business. Keep the change." He departed.

Stone and Dino both took note of his behavior. A few minutes later the woman returned and sat down.

"I'm very much afraid you have been abandoned by the cad you came in with," Stone said.

"Well put," she said, glancing at the empty seat across from her.

"Have you had dinner yet?"

"I had thought I had that in my immediate future," she said, "but it was not to be. Did he say something about important business?"

Stone nodded. "I'm Stone Barrington, and this is my friend, Dino Bacchetti," he said. "If you'd like to join us for dinner, we promise to be harmless."

"How kind of you!" she said brightly, moving toward them, while the waiter made adjustments with the table.

"Are you often abandoned so rudely?" Stone asked.

"Only by that particular gentleman," she replied.

"May I ask your name?"

"I'm Matilda Martin," she said.

"I know you may not believe this," Stone said, "but my mother's name was Matilda."

"Was she from Massachusetts?"

"From Great Barrington."

"Then I believe you. Was she a painter?"

"She was."

"When I was growing up, wanting to be a painter, she was my role model."

"Did you realize your ambition of painting?"

"Yes, but only on weekends. I'm a personal financial adviser." She took a folded brochure from her purse and handed it to him. "The illustrations are my own, but at least I'm published."

Stone looked at them. "You should paint more and advise less," he said.

"Excuse me," Dino said, leaning in. "Is there an angry ex-husband and/or boyfriend looming over your existence?"

"How well you put it," she replied. "Of course, every girl worth her salt has disappointed a man or two."

"Just checking," Dino said. "I wanted to see how Stone's personal cliché meshed with yours."

"As I mentioned, you should paint more and advise less," Stone said again, ignoring Dino.

"I shall take that as high praise," she said, laughing.

"That was how it was intended."

"Excuse me again," Dino said, "but there's a side of beef waiting out back to be slaughtered and grilled, too rare. Anybody interested?"

They ordered.

"Is the cad who recently abandoned you of any importance in your life?" Stone asked.

"I had hopes for him, but they didn't last long."

"Has he ever behaved violently toward you?"

"No, but he has threatened to, when I used up too much of the conversational air between us."

"Then he is living down to my expectations," Stone said. They finished their dinner and declined dessert. "Matilda, would you like to come back to my house for a nightcap? Dino will offer police protection."

"Dino, are you a policeman?"

"He is the *uber* policeman, the commissioner."

"Goodness, I've never met a police commissioner."

"I'm pleased to be your first," Dino replied. "I should tell you that Stone only wants to show off his house. You'll be safe enough."

"I'll show you my Matilda Stones," Stone replied.

"In that case, I'd love a nightcap," she beamed.

Stone filled the ride home with stories of being partners with Dino, in their youth.

"Did you and Stone protect each other?"

"Somebody had to keep him alive," Dino explained.

CHAPTER 3

Stone was at his desk the following morning when his secretary, Joan, rapped on his door.

"Yes?"

"A walk-in, says Bill Eggers sent him."

Bill Eggers was Stone's managing partner at Woodman & Weld.

"Better send him in," Stone said.

The young man who entered his office was all too familiar from the night before, when he had abandoned Matilda Martin at Clarke's.

"My name is Trench Molder," the man said, not offering a hand but taking a seat, unasked.

"I'm happy to know your name, Mr. Molder. How is it you know Bill Eggers?"

"I don't," Molder replied, "but I thought his name might get me in here."

"And now that you have accomplished that goal, how can I help you?"

"Simple. Just stay away from Matilda Martin, and we will have no further business."

"I'm afraid we already have no further business," Stone said. "But I can tell you that neither do you and Ms. Martin have any further business."

"That's presumptuous of you," Molder said

"Not really, since it reflects the wishes of the young lady."

"I presume she slept here last night," Molder said.

"She did not," Stone said. "Where did you sleep?"

Molder apparently did not wish to say. "Wherever I wanted to," he managed, finally.

"Good. Now, will you kindly leave these premises?"

"Or what?"

"That's the last time I'll ask you politely."

"Who gives a shit?"

Stone saw Joan appear behind Molder, her Colt .45 half-concealed in her skirt.

Stone shook his head. "It won't be necessary to shoot him," he said. "He's leaving right now." Stone stood up and walked around his desk. Molder stood, turned, and stalked out.

"Aw, shucks," Joan said. "I was looking forward to it."

"Maybe later," Stone said, "if he returns."

"I'll count on it," Joan replied, then went back to her desk.

Stone's cell rang. Dino. "Hello?"

"I just had a strong feeling that you might need me."

"That's very psychic of you, but fortunately the threat you imagined has vanished."

"Someone about our Matilda?"

"Yes, the cad from last evening. He has suddenly developed a proprietary interest in her."

"Did you tell him that ship has sailed?"

"I didn't bother. I just threw him out."

"Literally?"

"I was ready to, but he sensed that and fled."

"Did Joan want to shoot him?"

"Of course, but I stayed her hand. No blood on the carpet, and all that."

"A pity."

"He'll be back," Stone said, "as soon as he finds somebody to do the dirty work for him."

"Well, then Joan will get another crack at him."

"No doubt."

"Can I interest you in some tickets to the policeman's ball?" Dino asked.

"You mean they're still using that dodge to raise drinking money?"

"No, it's now a charity."

"Put me down for two," Stone said.

"Good. I'll bill you the thousand dollars."

"You're getting five hundred dollars a ticket? That's a little steep, isn't it?"

"It includes a corned beef and cabbage dinner."

"So, it's the *Irish* policeman's ball?"

"Right. You want tickets to the Italian one, too?"

"Take the grand and do with it as you will. And before you ask, I am *not* actually attending either of them."

"Got it." Dino hung up.

Joan buzzed. "Matilda Martin on two."

Stone pressed the button. "Good morning."

"Good morning. I just wanted to thank you for rescuing me last night."

"You're welcome. The young man, Molder, called on me this morning and ordered me not to molest you further."

"Further? You haven't even started!"

"I didn't want to tell him that. He seemed to want it to be true."

"One point: Trench is a coward, but he can still be dangerous."

"I rather thought so. That means he'll hire somebody else to beat me up."

"I wouldn't put it past him."

"I'll try and be ready."

"He'll choose a time when he thinks you're vulnerable and he's not."

"Of course. Dinner tonight, here?"

"Love to."

"Six-thirty?"

"Perfect, then you can molest me further."

"I'll look forward to it."

CHAPTER 4

Trench Molder spent the afternoon at his athletic club, toning up. His trainer, Howard Keegan, a retired Marine known as Huff, helped him with the weights, then sat down next to Molder at his invitation. "Something I can do for you, Trench?"

"No, Huff, but there's *someone* you can do for me."

"It would be my pleasure. How dead do you want him?"

"Not dead, just crippled a bit. I don't want to have skinned knuckles if the police should take an interest."

"Understood. That's my work, not yours."

"Five hundred?"

"That's generous, especially if I enjoy myself."

"You can enjoy yourself as much as you like," Molder said.

"Who is he, and where do I find him?"

"His name is Stone Barrington. He's an uptown lawyer who's messing with a girl of mine. He lives in Turtle Bay."

"I know the area. Security?"

Molder handed him Barrington's business card. "I was there this morning, and if he has security, I couldn't find it. Still, I think it would be best to take him away from home. I asked around and he travels in a green Bentley and has a small man as his driver."

"You want the small one hurt, too?"

"Don't bother. Just put Barrington in a hospital for a few days."

"As you wish. Description?"

"Over six feet, fairly solid build. Seems to think he's bulletproof."

"They're the best kind," Huff said, "the ones who think they can't be hurt."

"Something else. He has a friend who's the police commissioner, so don't get caught doing it. You should wear a mask. I don't want you ID'd by some passerby."

"Regular haunts?"

"P. J. Clarke's, Patroon. I don't know if he has a club."

"No matter. I'll do my research. Is he likely to fight back?"

"If he's conscious," Molder said.

Huff laughed. "I'll take him out with the first punch, then work on him at my leisure."

"That sounds like the way to go."

"Is there a message to deliver?"

"You can just say, 'Compliments of a friend,' before you knock him out."

"You want your name mentioned?"

"No, but you can say that your visit is a message from Matilda."

"Is that a real name?"

"It is. A bit old-fashioned, but it suits her."

"Should she be present to witness the beating?"

"Why not."

"Would you like anything done to her?"

"Whatever you feel like," Molder said, grinning.

"She can be the punctuation mark on delivery."

"Don't mark her face. I want her presentable."

"It's your dime. If that's what you want, that's what you'll get."

"Pick your own methods," Molder said. "I'll leave an envelope in your box, as soon as I hear the work is done."

"Fine by me. Does Barrington have any skills I should know about? Karate or boxing?"

"Nah, he's a ladies' man."

"Well, I'll see that he doesn't have anything to work with for a while. Final question: Is he likely to be armed?"

"I don't see any reason why he should be. He won't be expecting you."

"I'll do my due diligence, then." Huff left the locker room happy. He could use the unexpected five hundred.

CHAPTER 5

The next day Stone had a business meeting uptown with a client, and when he came back to the car, he saw a man looking at him. The man quickly looked away, and Stone filed the memory of the shape and heft of the figure as he crossed the street and strode away. He thought the man very confident, sure of himself. He wondered why the man felt that way.

Stone lunched with Dino at their mutual club in the East Sixties, a place so anonymous that it didn't have a proper name. Its members referred to it as "the club" or "the place uptown," and the membership was thick with men and women of influence.

"A threat?" Dino asked.

"Not exactly, but he looked as though he could be, if he chose to."

"Description?"

"Six three or four, wearing a blue suit. Thick neck, long arms, mostly bald."

"Did you get a look at his face?"

"No, he was more of just a shape. It was hulking, I suppose you'd say."

"I didn't know you were such a quick judge, Stone."

"When we leave here today, we'll see if he's hanging about."

"Okay."

After lunch, they left the club by its front door. Normally, they'd have driven away from the garage, which was more discreet, but Stone wanted a look around.

"There," he said, "just turning the corner."

"I only caught a glimpse," Dino said, "but your description was accurate."

"How would you classify him?" Stone asked.

"I think he's *capable* of being a threat," Dino said. "That's not necessarily an accurate assessment of his intentions, but it's troubling that you've seen him twice today already. I think you should treat him as a threat, until we know more. If you see him coming, cross the street. Then, let's see if he crosses with you. I don't suppose you're carrying."

"Maybe I should be," Stone replied.

"You'll live longer if you follow my advice," Dino said. "Carry a fucking gun."

"You think my guns are too light. I'm not going to carry a .45 just to make you happy."

"Then carry one of those dear little .380s you're so fond of. You're more interested in your suit hanging right than protecting your life."

"That's a dirty communist lie," Stone said. "I'm very fond of my life."

"Well, you can always shoot them in the head," Dino said. "Even a .380 will make an attacker think twice, if he can still think. Have you seen any more of that girl who has the same name as your mother?"

"Yes, and again tonight," Stone said. "Herbie Fisher is throwing a soiree, and we're going there."

"I got that invitation, too, and Viv gets back today, so maybe we'll see you there."

They parted. Stone got into the Bentley, which Fred, his driver and factotum, had pulled up beside them. Dino got into his own, city-supplied vehicle. As he was driven away, Stone heard a growl from the police car's siren, which made him laugh.

"Home, Fred," Stone said, "and I'll need you at six-thirty this evening."

"Yes, sir," Fred replied.

"Fred, have you seen anyone suspicious hanging about?"

"You mean the big fellow?"

"That's the one. I've seen him twice today, and that's too many times."

"I'll keep a sharp eye out," Fred said, and Stone knew he meant it. He also knew that Fred habitually packed.

"Will we be picking up a young lady, sir?" Fred asked.

"She's coming to the house, directly from her work."

"Got it, sir. The car will be ready at six-thirty."

Matilda was a little early, so they had time for a drink before departing for Herb Fisher's party. Standing in his study, she unpacked her large handbag, took off her dress, and liberated a black sheath from the bag.

Stone watched with pleasure.

CHAPTER 6

Trench Molder snapped the padlock on his locker shut, then walked into the weight room at the East Side Athletic Club and immediately saw Huff across the room, helping another member. Huff saw him, too, and quickly looked away. Trench didn't like that. Huff had no reason to be furtive with him.

Trench worked his way slowly through his routine. Eventually, Huff got around to him. "Hey, Huff."

"Hey, Trench."

"What's wrong?"

Huff heaved a deep sigh.

"Did you make any progress on the Barrington matter?"

"I'm sorry, Trench. Barrington made me, and so did his pal, Bacchetti."

"C'mon, Huff, you're slicker than that."

"I usually am," Huff said, "but not today."

"What happened?"

"I walked into his line of sight."

"So?"

"Twice. Now he's curious, and I can't approach if I'm on his mind."

"It could happen to anybody, Huff."

"Not to me. Not ever before. He's smarter than I thought."

"Smarter than I thought, too," Trench said. "So now, what's your plan?"

"Cool it down for a while. Otherwise, when he sees me on the street again, he'll approach, and then I'm useless to you."

"Cool it for how long?"

"Say a couple of weeks. I'll stick close to home. No contact at all. He'll eventually forget about me, and then I can do my work. I know you wanted immediate action, but at least you're not out any money."

"Let's leave it like this, Huff," Trench said. "You do your work in your own good time, and I'll be satisfied with that. On the other hand, I won't forget, either."

"That's good of you," Huff said.

"Just let me know when you're ready."

The party at Herb Fisher's was good, and Stone and Matilda immediately fell into the swing of things. There was a jazz pianist and a bass player, which kept the party moving. The noise level, as it does in a good room, went up.

Then, an unwelcome sight greeted Stone: Trench Molder walked into the room with a beautiful blonde on his arm.

"Did you see?" Matilda asked.

"I saw. Ignore him. If he speaks to you, don't respond. That'll give his girlfriend something to ask him about, and he won't have a good answer."

"Good idea."

They dragged some stools up to the piano and hung there. Stone liked the way the pianist played, putting thought into the music. Normally, as a kind of joke, Stone would ask a pianist to play "Lush Life," a Billy Strayhorn ballad so complex that nobody could remember it without the sheet music.

But this guy probably would, Stone surmised.

Herb Fisher greeted the two new guests, but he didn't look all that glad to see them, and he quickly moved on to speak to others. Molder looked unhappy about that, and his girl looked curiously at him. He took her elbow and moved her toward the piano, but Stone and Matilda both turned away as he approached, leaving his date with more questions to ask. Molder guided her angrily to the bar.

"Perfect," Matilda said. "Did you see his face?"

"I saw *her* face," Stone said.

Dino and Viv arrived and came over to the piano. Kisses and an introduction between Matilda and Viv followed.

"I see your cad of the other evening," Dino said to her.

"That's why he's here," she said. "To be seen. He'd like to be thought of as part of this crowd."

"Did you see anything more of your tail?" Dino asked Stone.

"Nope. He found somebody else to bug, I guess."

"I have a feeling you may see him again," Dino said, "and it won't be pretty."

"You could be right."

"Are you packing?"

"Not exactly," Stone replied.

"That's what I thought. I'll be scraping you off the sidewalk somewhere soon."

"I perceive no threats here," Stone said. "Not even from the cad."

Stone explained to Matilda what they were talking about.

"Describe the man," she said.

Stone did so.

"That sounds like a man who works at Trench's gym," she said. "Trench calls him 'Huff.'"

"You've met this Huff?" Stone asked.

"I picked Trench up at his gym once," she said. "He was talking to Huff when I walked in. There was a perfunctory introduction."

"I'm betting a perfunctory introduction is the only kind you'd want with a guy like Huff," Dino said.

CHAPTER 7

Trench Molder stepped out of the living room in Herb Fisher's apartment and into the guest room where coats were piled on the bed. He called a number on his burner phone.

"Yes?" Huff said.

"I'm at a party in a building a few blocks from you," Trench said, giving him the address and location. "Barrington is here with Matilda, and it looks like they're dug in for a while. I was thinking you might not need to wait a few weeks after all. What say you?"

"I can be there in twenty minutes."

"The green Bentley with the driver is parked across the street from the front door. You might have to deal with that, too."

"I can handle it," Huff said.

"Hurry up, then." Trench hung up his burner and returned to the party. The room had heated up, and someone had opened a sliding door to the large terrace on the front of the building. Trench took his date's elbow and steered her outside. "Let's get some air."

They walked to the front of the terrace and sat down on a padded bench next to the parapet.

"It's chilly out here," she complained.

Trench, annoyed, shucked off his jacket and handed it to her.

She put it on, then pointed at his shoulder holster. "Why are you wearing a gun?"

"Because I don't like being at the mercy of others," Trench replied, "and I'd rather defend myself up front than wait for somebody else to make the first move." He leaned over the parapet and saw the waiting Bentley. "You'd be surprised if all the men at this party removed their jackets."

"I didn't know paranoia ran that deep," she said.

Trench resisted the temptation to slap her across the face. He didn't need a scene that would cement his location at this hour into other guests' minds.

They had been sitting there for half an hour before Trench saw Barrington and Matilda make a move inside. Barrington went to the spare bedroom and came back with their coats. Then they said their goodbyes to their hosts and walked to the elevator.

Trench pressed the button for the last number dialed on his throwaway.

"Where are you?"

"Out front."

"They've just left the party and are on the way down."

"I'm on it."

Trench looked at his date and saw her reaction to the call, but she said nothing. He stood and rested his elbows on the parapet, looking down at the street. The Bentley was nicely illuminated by a streetlamp. He saw a movement in the doorway of the building across the street. Huff was ready.

"Can we go back inside now?" the girl asked. "I need to stand by the fire and warm my ass."

"Don't worry," Trench said. "I'll warm it for you in a few minutes."

She stood up and looked into the street. "What are you looking at?"

"There's a Bentley down there."

"Yes, I saw it when we got out of the cab."

"I'm thinking of ordering one."

"Based on how it looks from ten stories up?"

Trench tensed. Barrington had just walked out of the building. The driver got out to open the door on the curbside for them.

"Okay," Trench said, "let's go back inside and warm your ass." He took his jacket back and put it on. The police would be here soon, and he didn't want to have to explain the shoulder holster to them.

CHAPTER 8

Stone approached the Bentley with Matilda on his arm and motioned for Fred to get back in the car. He put Matilda into the back seat and walked around the car to enter from the other side. He opened the car door and lifted a foot to enter. As he did, he saw a blur of motion in the corner of his left eye, and something solid struck him across the back of his neck. As he descended into unconsciousness, he heard what he thought was a single gunshot, which echoed in his head.

The echo was the first thing Stone heard when he opened his eyes. He took a moment to orient himself, then it became clear that he was in a hospital room, and a rather nice one. There was an IV running in one arm, and a vase of beautiful flowers on a table by the door.

"His eyes are open," a woman's voice said.

Stone focused on the face hovering above him, and as her features became sharp, her name did, too. "Matilda?" he said. "Mom?"

"He's awake, and he thinks I'm his mother," she said to someone.

"Only joking," Stone replied. "Forgive me for asking the obvious question, but: Where am I? Is this a hospital or a mortuary?"

"It's Lenox Hill Hospital," Dino's voice said. "And stop being a wiseass."

"But it's so much fun," Stone said, "and it helps with the headache. You got any aspirin on you?"

"Oh, shut up. I liked you better unconscious. Here's the nurse."

"Mr. Barrington?"

"Yes, but you can call me Stone."

"What do you need?"

"Something for my headache and my sore neck."

She produced a syringe and injected the contents into his IV bag, near the bottom. "You'll feel better in a few seconds."

She was right. He was flooded with warmth, and the pain flew away. "Okay, Dino, what the fuck happened?"

"Somebody used a blackjack on you, and he was an expert."

"Anybody we know?"

"Remember Huff?"

"Oh, yeah. Where is he?"

"Last time I saw him, he was lying dead on Park Avenue."

"How'd he get that way?"

"Fred put a bullet into his chest—dead center, one shot."

"And where is Fred?"

"Downstairs with the Bentley. I cleaned up after him, and he hasn't been and won't be charged with anything. You'll have to give him another gun, though. His old one is in police custody."

"How many times did he have to shoot Huff before it worked?"

"It worked the first time. Huff is now a part of your storied history."

"Matilda, are you all right?"

"That's the right question," she said. "I am."

"I'm glad to hear it. Did Huff hit you with a blackjack?"

"No, I was too busy dealing with the unconscious man in my lap."

"How long have I been out?"

"About three hours," a male voice said, stepping up from behind the group. "I'm Dr. Herbert. How do you feel?"

"Just fine, and I will be so until you run out of morphine."

"You've got yourself a concussion," Dr. Herbert said. "And you're going to have to stay in bed another day, until we decide you're not going to die. Leave the morphine to me."

"Okay," Dino said. "I guess I can cancel the appointment with the undertaker. He's going to be terribly disappointed. He wanted to give you an inspector's funeral."

"Tell him I'm not very disappointed. Next time, maybe. Can somebody crank this bed up to a sitting position, so I'll feel less like a fresh corpse?"

Somebody found a button and held it down until Stone was satisfied with his position.

Stone looked around. "Everybody's here," he said, "except Bob. Where's my dog?"

"We had to restrain him from visiting," Joan said from across the room. "He'll be waiting when you come home."

"Tell him '*woof*' for me."

"Done."

"If I promise not to die, will you let me out of here now?" he asked the doctor.

"You're in no condition not to die. If you stood up now, you'd fall down."

"Oh."

"You know, you are the first blackjack victim I've had for years. I'd forgotten how effective those things can be."

"I would arrest Huff for assault with a deadly weapon, if he weren't already dead," Dino said. He tossed a blackjack into Stone's lap. "There's a souvenir for you. Huff's is in an evidence bag, but it's the same kind as this one."

"And you just happened to have it lying around?" Stone asked.

"A good cop never reveals his sources."

"May I?" the doctor asked, picking up the thing and slapping it into his palm. "Oh, to be a cop."

"Cops aren't allowed to carry blackjacks anymore," Dino said. "It was too easy to kill somebody with one."

Dr. Herbert set the blackjack down, said, "See you later," and walked out of the room.

"I'm going to get some sleep," Stone said, and then he was unconscious again.

CHAPTER 9

Stone was home in time for lunch the next day. Bob greeted him enthusiastically, and Stone found him a cookie to reward his adoring opinion of his master.

"Want some good news?" Joan asked.

"Always," Stone replied.

"I'm closing on my house soon." Joan had inherited a very large town house off Fifth Avenue, and it had been too much for her.

"What did you end up getting for it?"

"Twenty-three and a half million, mostly furnished. I took some pictures and a few pieces of furniture."

"Good move. Where will you live now?"

"Back in my apartment next door. It's all the house I need for me and, when you're gone, Bob."

Bob wagged all over at the mention of his name, and Stone scratched his back.

Dino walked into the house. "I came to see if you were dead yet."

"Sorry to disappoint you."

"I'm not disappointed; I'm elated to see you breathing on your own."

"Well, if you'll forgive me, I'm going to lie down for a few minutes." He stretched out on the sofa. "Have you arrested Trench what's-his-name yet?"

"On what charge? Knowing a guy who hit you with a blackjack?"

"Isn't that a crime?"

"It is to you, of course. But a sober judge would take a different view, and the DA would yell at me for hauling him in."

"Where's Matilda?"

"She stopped by her place to pick up some clean clothes," Joan said. "I expect she'll be along soon."

"We need to find out absolutely everything she knows about Trench," Stone said.

"You want to buy him a present?" Dino asked.

"I want to buy him a prison term."

"Well, you're going to have to wait until he does something provable to you."

"I would have thought that blackjacking me was probable cause."

"It would be, if we could prove he was involved. Fred erased the only guy who could have testified against him."

"Was Fred, ah, a little hasty in his judgment?"

"No, Huff's hand—the one without the blackjack—held a .25 automatic, just the thing for putting your lights out permanently, if the blackjack failed to operate as intended."

"Well, at least Fred is free and clear. Has he still got his carry license?"

"I saw to that."

"Good, in case he has to shoot somebody else. Did Huff have a sheet?"

"A couple of bar fights. I guess he was staying in practice."

"How about Trench?"

"About two hundred parking tickets. I handed that off to a guy in the DA's office who makes life hell for people who tear up their tickets."

"I'd like to see him in something more permanent than parking ticket hell."

"Maybe he'll give us another shot at him," Dino said. "Of course, that would mean he'd have to take another shot at you first."

"I know, I know, you want me to carry all the time."

"Everywhere but on the tennis court."

"They've probably got a holster made for that. It would be intimidating to your opponents, too. Nobody would want to play with you."

Stone's secure cell phone rang.

"One guess who that is," Dino said.

Stone pressed the button. "Hello, Lance." Lance Cabot

was director of the Central Intelligence Agency, for whom Stone and Dino performed consulting duties.

"I'm delighted to hear that you can still speak," Lance said, not sounding delighted.

"Don't worry, it wasn't a matter of national security."

"Are you quite sure about that?"

"I don't have any reason to believe that was the case."

"You're pretty hopeful for a man who's just spent a night in the hospital."

"It was sort of a high school playground fight that got out of hand."

"Really? I heard your assailant was a professional hit man."

"If he was, the NYPD doesn't know about it."

"There is a great deal that they don't know."

"Not so loud. Dino will hear you and take offense."

"Put him on, and I'll offend him directly."

"I'm going to take a nap now," Stone said.

"Sweet dreams." Lance hung up.

CHAPTER 10

Matilda turned up as Stone was waking from his nap. Dino was still there, making an occasional call to his office.

"Aha! Matilda!" Dino said, loudly enough to bring Stone to full consciousness.

"Dino! Stone!" she cried, imitating Dino's enthusiasm.

"Sit down, Matilda," Dino said. "We want to interrogate you."

"On what subject?" Matilda asked, looking alarmed.

"On the subject of Trench Molder," Stone said, sitting up. "We need to know everything about him, especially how often he hires assassins to kill people he doesn't like, such as my own self."

"Okay, let's see," Matilda said. "I met him at a party at somebody's house on the Upper East Side."

"Whose party?"

"I don't remember the name. A girlfriend had been invited, and she took me along. I do remember the host had a grand piano, because somebody was playing it, and somebody else was singing."

"Was either of them Trench?"

"No. Trench was sitting on a stool at the piano, next to me."

"Doing what?"

"*Trying* to sing along, until the piano player told him he couldn't carry a tune and to shut up."

"Did he have the pianist beaten up or killed?"

"I don't know. Maybe, since I never saw him again."

"What were your first impressions of Trench?"

"A little drunk, funny, cute."

"Did he introduce himself?"

"He did. And he gave me a card, which I have since discarded, before you ask for it."

"What kind of work does he do?"

"I believe work is what he would call a four-letter word. He lives the life of luxury, thanks to a relative. An uncle, I think."

"Do you know this uncle's name?"

"He never mentioned it."

"Was there anything threatening or sinister about Trench?"

"No, except that once he learned my name, he seemed to think he had taken possession of me. If another man tried to enter our conversation, he became surly. That proved to be a reliable indicator of his future attitude toward me."

"Did he ever mention having committed a crime?" Dino asked.

"He mentioned having beaten up a couple of people who intruded on his company with a woman."

"Did he say if he hired someone to do that for him?"

"Sort of, but I don't remember what words he used."

"Did it surprise you to learn, when you learned, that he would hire someone for that purpose?"

"It surprised me that the man he hired had a gun, which indicated that he planned to shoot somebody, probably Stone or me."

"Was Trench mad enough at you to have you killed?"

"I don't know. I couldn't see inside his head."

"Did he mention ever having done that before?"

"He managed to give me the impression that he did not suffer competition gladly."

"And Stone was the competition?"

"That's my guess, given what ensued."

"Was Trench watching what happened?"

"He was still upstairs at Herb Fisher's party, so there is a chance he could have been looking down on us. I didn't catch him doing it. I was already inside the car when the trouble started."

"After the police arrived and people were swarming around the crime scene, did you see Trench among them?"

"No, I didn't get out of the car until we reached Lenox Hill."

"Did you see Trench hanging around Lenox Hill?"

"No, but I wasn't looking for him."

Dino paused in his questioning.

"Dino," Matilda said. "I think you've pumped me dry on the subject of Trench. Is there something else you'd like to question me about?"

"Ah, no," Dino said. "I grant you bail. But I may want to talk to you more later."

"Gee, I hope I can make the bail."

"You'll have to forgive Dino," Stone said. "When he gets excited about an interrogation, he tends to hyperventilate and has to breathe into a paper bag for a while. He's almost at that point now."

"I remain at your beck and call, Commissioner," Matilda said.

"That's how I like my witnesses," Dino said. "No paper bag necessary."

CHAPTER 11

Trench Molder walked into the weight room of the East Side Athletic Club and looked around. He hoped to give the appearance that he was looking for Huff to help him with the weights. Huff's assistant manager, known as Bozo, approached.

"Good afternoon, Mr. Molder."

"Good afternoon, Bozo. Is Huff off today?"

"I take it you have not watched the morning news or read the *Post*."

"You are correct, but why are you concerned with my viewing and reading habits?"

"Because, if you had been viewing—or reading, for that matter—you would know that Huff was shot dead on Park Avenue last evening."

Trench's eyes widened. "Surely, you jest."

"I do not," Bozo said. "All we know is that he seems to have hit someone with a blackjack—I know he had one, because he showed it to me, once—and the man's driver pulled a gun and shot him dead."

Trench sat down heavily on the bench and put the back of a hand to his forehead.

Bozo lowered his voice, and said, "I know, I know. This was nothing to do with the club. This was freelance work that Huff sometimes took on. I assisted him, once or twice."

"Was his killer arrested?" Trench asked.

"Apparently not. According to the *Post*, he was questioned at the scene, then released to drive his employer—who had suffered the blackjack attack—to the ER. Then he was questioned again and released again. The police judgment was that the shooting was legal and justified, in the circumstances."

"Who was the man Huff attacked?"

"Someone named Barrington, a lawyer."

"I've heard the name."

"Apparently, Huff knew him well enough to hate him."

"Do you know why?"

"No, sir, I don't." Bozo looked around to be sure no one was within earshot. "I would like you to know, though, that should you require the kind of assistance you once received from Huff, I would be glad to step into the breach."

"Good to know, Bozo," Trench said. "I think I'll skip my workout today and just hit the showers."

"I understand, sir," Bozo said.

Trench left the club, freshly showered, but still pretending to be upset about Huff's fate. He was, in fact, upset. How could Huff get close enough to the man to blackjack him, but not to finish the job? It didn't make any sense. He made a note to himself that Barrington's driver was armed and dangerous to anyone who approached.

Trench walked the few blocks to his apartment building. As soon as he was in the elevator, he shook off his pretense of mourning and assumed his normal mien. With Huff gone, he needed new help. He wondered if Bozo was a reliable person. He sat down in his study and made a call to a little man named Joe Rouche, who did errands for him.

"Good morning, Trench," Joe said. "What can I do you for?"

"I want a thorough check on a man who works at my gym. He's called Bozo. I don't know his proper name."

"I'm on it," Joe said. "You want it in writing, or just oral?"

"Oral will do," Trench replied. "I just need to know if he's a reliable man. If I can trust him with, ah, work of a confidential nature."

"I'll get back to you," Joe said, then hung up.

Joe already knew who Bozo was, and something about his character, which he would describe as dubious. Still, he used his computer to do some research, taking written notes as he worked. He already knew about Huff, too. He imagined that the man had been on an assignment from Trench Molder when he met his fate, and now Molder wished to replace him.

He called Bozo.

"Hey, Joe," the man said.

"Hey, Bozo. I've had a request from an old client to take a look at you and report back. What would you like me to say?"

"I think I know who you mean," Bozo replied. "Huff was probably working for him when he got himself plugged last night. He's looking for better help, and that's me."

"How much do you want me to tell him?"

"Tell him whatever you've got. Don't lie to him because he might have somebody else running another check. I'll be suitably grateful to you for a positive recommendation."

"Okay. I'll report in and leave the 'suitably grateful' to you, as long as you treat me right."

"You can rely on me for that," Bozo said.

The two men hung up.

CHAPTER 12

Stone took longer to fully recover from his blow to the head than he had imagined he would. After three days of being dizzy and unsteady on his feet, he called the doctor, who told him to stay off his feet for the rest of the week, unless he used a cane or a walker. Stone flinched at the mention of those two implements. He was also told not to go out for dinner or any other reason.

"It's this way, Stone," the doctor said. "If you start strolling around too soon, you'll fall down. And a fall could do more damage than the blackjack did. Just be a happy invalid until you're steady again."

Stone found that his libido suffered, too. Or, rather, Matilda discovered that fact. "You've gone off me," she said, when he failed to salute on demand.

"No, no, no," Stone said quickly. "I've gone off my

rocker, but it is only temporarily. I had a talk with my doctor today, and he told me to stay off my feet."

"What about your back? Are you to stay off that, too?"

"I'm not supposed to exert myself, and you are a walking, talking exertion." He wiped sweat from his forehead. "You see this? It's not hot in here, but I'm sweating from the effort of just chatting."

"You want me to leave you alone, then?"

"No. You could give me a back rub, though."

"I was a masseuse in my extreme youth," she said.

"There's a table in my dressing room." He pointed.

Soon she was kneading away.

"Not my neck," he said. "That's still too sore."

"Turn over on your back," she said, half an hour later.

Stone managed to get flipped, but he held a pillow over his crotch.

"What's the matter, is it cold?" she asked.

"It's cool, and I want it to stay that way. I'll owe you."

"You're going to get so deeply in debt, you'll never get out from under," she said, pinching a nipple.

"You stay away from what a friend of mine used to call, 'the erroneous zones.'"

She laughed. "Oh, all right. I'll wait until you're fully recovered, then look out!"

"I'll look forward," he said.

Dino called in the late afternoon. "Dinner at Clarke's?"

"No, I'm under doctor's orders not to go out. He's afraid I'll fall down and hurt myself."

"That's always a consideration when you drink that hundred-proof bourbon you like so much."

"Come over here for dinner, and bring Viv."

"All right, see you then."

Joe Rouche called Trench Molder.

"What have you got for me?" Trench asked.

"Nothing too alarming. Thomas David Bozeman, thirty-two years old, very physically fit—not unusual, given his work. A couple of arrests, one of them for clocking a woman in a restaurant, the other for winning a bar fight. He's thought, by those who know him best, to be reliable and a pretty good guy. The athletic club promoted him to Huff's job, and they wouldn't have done that if he hadn't been okay. Their members don't like trash working there. There's been some grumbling about Huff's use of the blackjack, and Bozo won't have that. Generally speaking, he does what he says he'll do, so he's viewed as trustworthy by those who know him."

"Okay," Trench said, "good to know. My man will send you a check." He hung up and looked at his calendar. Too soon for another shot at Barrington, he thought. He would wait until the memory of Huff's blackjack faded.

Viv held Stone's chin in her hands and looked into his eyes. "You're not well," she said. "Not yet."

"Funny, that's what the doctor says, too, and it's why we're dining in."

He started to rise, but she pushed him back into the sofa. "I'll get the drinks. You just sit there and look pretty."

"Whatever you say," Stone replied.

Dino looked at him closely, too. "She's right," he said.

"She usually is," Stone agreed. "Look, I'm taking her advice."

"That's a smart move. Otherwise, she'll beat you up."

CHAPTER 13

Stone nodded off over brandy and had to be taken upstairs by Dino, using the elevator. There he delivered Stone to the tender mercies of Matilda, who got him into a nightshirt, his favorite mode of dress at bedtime when he wasn't naked for a good reason.

Matilda slept next to him, checking his inert form every time she rolled over. He was always the same.

Stone woke, rested and clearheaded for the first time since suffering the blow. Matilda was nowhere to be found. He had breakfast in bed, read the *Times*, and did the crossword. Finally, he got to his feet and walked around without falling down or bumping into anything. He shaved and took a shower without either cutting his throat or slipping on the wet tiles. Finally, he got dressed and walked downstairs to his office, just as if he were perfectly well.

"You look perfectly well," Joan said, when she came to inspect him.

"Thank you, that's exactly how I feel, just don't hit me over the head with anything."

"I'll try to avoid that." His phone rang, and Joan answered it at his desk. "Surprise," she said, "Dino on one."

Stone picked up. "Good morning."

"You sound—I don't know, normal," Dino said.

"And that's how I feel."

"Dinner tonight, Clarke's?"

"See you then. Viv?"

"Maybe. Matilda?"

"Maybe. I haven't seen her this morning."

"Whatever." Dino hung up.

"Matilda went back to her place," Joan said. "She didn't say why or for how long."

"Perhaps I've been inattentive," Stone said.

"Perhaps." The phone rang. Stone picked it up before Joan could. "Hello?"

"You sound normal," Matilda said.

"I *am* normal," he replied.

"I'm doing laundry."

"You could have done it here."

"I needed fresh clothes. You were getting tired of the ones I was wearing."

"Your imagination," he said. "Dinner tonight at Clarke's?"

"Where we met! How romantic!"

"Shall I pick you up?"

"I'll meet you there. I've got a previous engagement with Bloomie's."

"I wouldn't want to come between you two. Seven?"

"Good." Everybody hung up.

"Well," Joan said, "I guess you're officially normal."

They assembled in the bar at P. J. Clarke's, had a drink, then went to their table.

"Did you notice the man in the blue blazer a couple of seats down the bar from me?" Matilda asked.

"I didn't," Stone replied.

"I did," Viv said. "Who is he?"

"I don't know his real name, but he's called Bozo. He worked at that athletic club for Huff, whom you will recall."

"Actually, I recall nothing about Huff. First, he was standing behind me. After that, he was dead, and I was unconscious."

"You'll have to trust me, then," Matilda said. "Bozo was Huff's assistant. I knew about them both from Trench."

"He looks well-muscled under the blazer," Viv said.

"Trust you to notice that," Dino interjected.

"Trench admires people who can fight," Matilda said. "And, after all, Bozo does work in a gym."

Dino looked at Stone. "Are you packing?"

Stone sucked his teeth for a moment. "I forgot," he said.

"Well, Viv and I both are, so we'll shoot anybody who bothers you. After this, though, you're on your own."

"I'll try and remember that," Stone said.

"I'll remind him," Matilda said.

"Are you packing, Matilda?" Dino asked.

"My bags or my gun?"

"Do you own a gun?"

"Trench left one at my apartment once and never came to get it."

"Are you licensed?"

"No."

"Do you know how to shoot a gun?"

"Yes, my father taught me."

"I'll send you the license application. Don't leave your house in possession of a weapon until you've got the license in hand. I'll expedite it."

"Thank you, Dino."

"Also, I don't think you should keep the gun Trench left at your apartment. You don't know where it's been or what it's done."

"I'll send it back to him, then."

"Unload it first, and put the cartridges into a separate bag. Don't mail it. Send your doorman to his place with it securely wrapped. You want it off your hands as soon as possible."

"Why don't I have Fred pick it up from your place and deliver it?" Stone said.

"Everybody is so solicitous," she said.

"We have to be," Dino replied. "When you have a gun."

CHAPTER 14

Matilda's carry license turned up, hand-delivered, two days later, and Stone took her downtown to a gun store frequented by cops.

"This is a wonderland of mayhem," she said, looking around at the showcases and wall display of weapons.

"You need to choose only one," Stone said, leading her to a showcase. "Let me see your hand."

She spread her fingers.

"Not unusually long," he said. "Let's try gripping this one." He pointed at a Sig Sauer .380. A salesman unlocked the case, cleared the weapon, and handed it to her.

She held it in front of her and look down the barrel. "It's a nice size," she said, "but it's a little too heavy."

Stone turned to the salesman. "Do you have a Colt Government .380?"

"Colt doesn't make it anymore, but I've got a couple

of nice used ones." He went to another case and came back with two small pistols, which were, to the eye, like miniature Colt Model 1911 .45s. One was nickel-plated.

"The other one," she said, pointing at the blued pistol. The salesman cleared it and handed it to her; she hefted it in her hand and pointed it at a target across the room. "Perfect."

Stone paid for the weapon and a box of cartridges, the salesman put it in a box, and they left. Back at his house, they went down to his one-lane shooting station in the basement, and she fired some rounds. Soon, she was hitting the paper outline of a man in the chest, with a nice grouping.

Stone showed her how to disassemble the weapon and clean it. "Must be done every time you fire it." He wiped it clean with a cloth and dropped it into the zippered leather envelope that the pistol had come in. "Okay. You're officially dangerous."

They went back upstairs, where he gave her the license, and she tucked it into her purse.

"Always carry it," Stone said.

Fred knocked and entered the room. "I delivered the pistol to the address indicated, and the doorman signed for it," he said, handing Matilda a receipt.

"Hang on to that," Stone said. "Now, if Trench uses the pistol in a crime, you can prove you were not in possession of it."

"I thought Ms. Matilda might like to see this," Fred said, handing her a catalog. It was full of holsters and other carry items.

"You can strap it to your thigh or hide it in your Thunderwear," Stone said, pointing at the item.

"Thunderwear," she said. "I like the name."

"Or you can hide it in this bra, although yours is already pretty fully packed. There are various handbags designed to conceal the gun, too."

"I'll decide later," she said, putting the catalog in her handbag.

"Now, having done all that, I have to give you a stern lecture," Stone said.

"I'll brace myself."

"Now that you are fully licensed and armed, there's something you must never do."

"What is that?"

"Never shoot anybody."

"What? What's the gun for then?"

"To keep you from having to shoot anybody."

"I don't understand."

"If you shoot someone, even if you don't kill him, your life will change forever, and not for the good."

"How so?"

"You will find yourself suddenly surrounded by policemen and technicians, all trying to prove you have broken the law."

"Even if I haven't?"

"That's how they prove that you haven't broken the law," Stone said, "by trying and failing to prove that you have. Your name will appear in a nationwide computer database of people who have been investigated for a weapons charge, and sometimes they forget to mention that you were innocent of any crime. Every time someone is shot with a weapon of your caliber, your name will be printed out on a list of people who own one, and you will be investigated all over again. They will confiscate your pistol and test it against the available evidence, and you will be without it for weeks, perhaps months. Then the legal stuff will begin."

"What legal stuff?"

"The person you shot will sue you for grievous bodily harm, and the costs of being sued will mount rapidly—again, never mind that you have done no wrong. There will be lawyers out there who find you a tempting target. Even if you win your case, everyone will know that you were charged, and they will forget that you were innocent. Are you getting the picture?"

"Never shoot anybody," she said, solemnly.

"Engrave that on your frontal lobe."

CHAPTER 15

Trench Molder received the package from the doorman as he walked into his building. He took it upstairs and unwrapped it.

"Shit," he said. "What is she up to?" He picked up the phone and called Matilda.

"Hello?"

"Hello, yourself. It's Trench."

"I wasn't expecting your call."

"Why did you send the gun back to me?"

"It belongs to you."

"I gave it to you."

"I don't want it."

"Why not?"

"Because I don't know where it's been or who it has been used to kill."

"It's brand-new; a virgin weapon."

"I have obtained another weapon, and a license to use it."

"They gave you a carry license? It took months for me to get one, and cost ten thousand dollars."

"Funny, mine didn't cost anything."

"How did you do that?"

"I filled out a form and got fingerprinted. I didn't bribe anybody."

"Is this conversation being recorded?" he demanded.

"Only if you're recording it. And let me say this for the record: don't call me again, because next time you do it will be recorded, if you incriminate yourself."

"You bloody bitch!"

"I don't know about the bloody part. But I'm happy for you to consider me to be a bitch, because that's what I will be, if you reach out to me again."

She hung up. "How'd I do?"

"You made your point," Stone said. "I wish we could have learned who he paid ten thousand dollars to get the carry permit for him. I would have liked for Dino to nail that bastard."

"How much did you pay Dino to get mine?"

"Listen to me very carefully," Stone said. "All Dino did was to move you up in the line and write a letter swearing to your good character. You're ungrateful to insinuate that he broke the law by doing you a favor."

She looked away, abashed. "I'm sorry, you're right. I shouldn't have said that."

"Just as long as you know the truth. Tell me, did you ever have occasion to sniff the barrel of the pistol Trench gave you?"

"You mean, like, take cocaine off it?"

"No, I mean to ask what it smelled like, if you smelled it."

"I didn't smell it. I didn't know I should have."

"It was not a requirement, just a precaution."

"Against what?"

"If the gun smelled of oil, it would not have been fired recently. If it smelled of gunpowder, then it would have been fired."

"What does gunpowder smell like?"

Stone thought about that. "Like gunpowder, I guess. I don't know what compares."

"Now am I safe from being accused of a crime?"

"No, but nobody can prove that you committed one—not with that pistol, anyway. Do you think that Trench is capable of using it to murder someone, then giving it to you?"

That stopped her. "I don't know, but after his attitude in that conversation, I wouldn't be surprised."

"Give me the receipt that Fred gave you, and I'll put it in my safe."

She dug it out of her handbag and handed it over, and he put it into an envelope, wrote her name on it, and put it into his safe. "There," he said, "that problem solved."

She sat down in his lap and snuggled. "Are you going to solve all my problems?"

"That depends on what problems you have. If you want to tell all, go ahead, and I'll do what I can."

"Let me think about it," she said.

"That bitch."

Trench wanted to throw his phone against the wall but didn't give in to the urge. That asshole attorney must have told Matilda to send the gun back.

He called Bozo.

"Hi, Trench."

"Are you still interested in a little side work?"

"I am."

"There's a problem I'd like dealt with posthaste."

"The lawyer you had me follow the other day? Barrington?"

"One and the same."

"You want me to finish what Huff fucked up."

"Very much. But not just Barrington. Matilda Martin, too."

"Are we talking about a don't-mess-with-me message?"

"No. Something more permanent, if you get my drift."

"I do. Give me a day or two to set things up."

"The sooner the better. And, Bozo, one more thing."

"Yes?"

"I want to be there when it happens."

"I'm not sure that's a good idea."

"It is if you want to get paid."

"On second thought, I'd love to have you along."

CHAPTER 16

Matilda went upstairs, and Joan buzzed Stone. "Carly Riggs on one for you."

Carly had been a graduating law student at Yale, number one in her class, when Stone first met her. He had been so impressed that he recommended her to Woodman & Weld for a job as an associate.

Stone pressed the button. "Carly? How are you?"

"I'm terrific," she replied, "as always."

She made him laugh, too. "How's life at Woodman & Weld?"

"Interesting. I took the bar exam this morning."

"So soon? I thought new associates spent a couple weeks cramming for it."

"Nah, I read the cram book once, and I think I aced it."

That she would say that out loud was typical Carly. She did not have the same social filters as others.

"When do you get the results?" he asked.

"They say in a week."

"I think it's likely that somebody at Woodman & Weld will whisper the results in your shell-like ear before that. Somebody told me when I had taken it that I'd finished in the ninety-something percentile, and only the next day."

Someone spoke to her at the other end. "Can I call you back?" she asked.

"Sure. The law comes first."

She hung up.

Stone's phone rang again. It was Carly.

"It's me again. You were right, somebody told me I did very, very well on the bar, but they didn't have a percentile."

"Then you've got nothing to sweat. Relax and learn how to practice law."

"Oh, I already know how to do that—the criminal part, I mean."

"Yes?"

"Sure, I've seen every episode of *Law & Order* at least twice, and I never forget anything."

"You'll find there's more than that to learn, and don't forget the civil side—that's where most of the money is made, year in and year out."

"Well, they've got me working for Herb Fisher, and he practices whatever law walks in the door, so a year or so of him, and I'll be ready for a partnership."

"Carly, take my advice: never even whisper those words to any person at Woodman & Weld, or you'll discover you have a big target on your back. You must try, as hard as it may be, to display a becoming amount of modesty, even though that is not in your nature."

"Okay, sure."

"I know it won't be easy for you, but know that it will make life easier for you in a big law firm. And handle Bill Eggers, in particular, with kid gloves. He's your managing partner, and your career rests in his hands, not yours."

"I'll remember that."

"Forget it at your peril. You're going to have to acquire some editorial skills when wagging your tongue. A slip can come back to haunt you, when you least expect it."

"I understand."

"I hope to God you do, or you'll find yourself hanging out a shingle at an office in the cheapest neighborhood in the city. Remember, there are two words that can damn any young lawyer to hell in a handbasket. The words are 'private practice.' At cocktail parties, especially ones where there are a lot of lawyers present, hearing those words will cause eyes to glaze over."

"You sound as if you want me to be afraid," Carly said.

"Just a little fear can go a long way to make life in the law easier. And don't talk to Herb Fisher the way you talk to me. He's in a position to cut you off at the knees, if you annoy him or insult his intelligence. Remember, he's almost as smart as you are, and he has a *lot* more political savvy—in office politics—so shut up, listen, and learn."

"Oh, all right."

"And no matter what you do, don't let him get you

into bed. He won't tell anyone, but everybody in the office will know immediately."

"If I don't tell anybody, and he doesn't tell anybody, how will they know?"

"They can smell it on the breeze," Stone said. "Trust me." He paused. "You haven't slept with him already, have you?"

"No, I haven't."

"But you were thinking about it, weren't you?"

"Oh, stop it."

"Here's how to handle this: pretend that everybody in the office thinks the two of you have slept together, so spend all your time convincing them you have not."

"That makes a weird kind of sense."

"Would you like to have dinner with Dino and me this evening? Bill Eggers will be joining us, too. It would be good for you to get some face time with him."

"Sure."

"You know P. J. Clarke's?"

"Of course."

"Meet us at the bar at six-thirty. Oh, and Dino's a fount of information on everything cop, so listen to him!"

"Got it."

They both hung up.

CHAPTER 17

Carly, characteristically, got to Clarke's first. She was perched on a barstool, showing a lot of leg, sipping a martini, when Stone walked in with Matilda.

The two women sized each other up as introductions were made.

"How long have you been here?" Stone asked Carly.

"Five, ten minutes."

"How many offers have you had?"

"About that many."

"I hope you turned down all of them."

"Well, not *all* of them. There were one or two acceptable applicants."

"Be careful."

"That's all you ever say to me."

"It's all the words I can get in edgewise when you're talking, which is all of the time."

"Stone," Matilda said. "I'm sure Carly can handle herself. I doubt this is the first time she's been approached."

"Oh, no," Carly said. "Not even close. Would you like an exact number?"

Matilda chuckled as if Carly were joking.

"She's not joking," Stone said, then turned to Carly. "Perhaps you should save that information for another time."

"If you think that's best."

"I do."

Dino joined them.

"Hi, Dino," Carly said.

"Hey, Carly. Congrats on your graduation, your new job, and whatever you got on the bar exam."

"I can see that Stone is keeping you up to date."

"It's his job," Dino said, "on those rare occasions when he knows more than I do about something—in this case, you."

"Where are you living?" Stone asked her.

"On a leather sofa in Murray Hill," she replied.

"I can do better than that. I've got an empty flat next door to my house, where my staff lives. There's even room for your BMW convertible."

"Oh, I negotiated for twenty-four-hour parking with Woodman & Weld," she said.

"Smart move."

"So, when can I make the smart move to your place?"

"Anytime you like."

"My suitcase is in the cloakroom, as I was about to jump couches this evening," she said. "How about after dinner?"

"That will be convenient."

"Then my life will be complete."

"You're even more generous than I thought," Matilda said to Stone. Though she smiled, there was no hiding her discomfort with the turn the conversation had taken.

"Only to those who've earned my generosity."

"Like me?" She wrapped her arm through his, staking her claim.

"That goes without saying."

"Please, do say it."

"Like you."

Bill Eggers walked in, and they used his arrival as an excuse to sit down at their table.

"What's new, Bill?" Stone asked.

Eggers looked uncomfortable. "This can't get out," Eggers said. "I mean, it'll be published next week, but it can't get out until then."

Stone thought Eggers looked about to bust. "You'll have to trust us," Stone said. "Otherwise, you'll explode."

Eggers flicked a glance at Matilda.

"Don't worry, Bill. I vouch for her character. Not a whisper will escape her lips."

Matilda mimed zipping her mouth closed.

Satisfied, Eggers turned to face Carly. "It's about you."

"Oh," she said. "You've found out already?"

"Found out about what?" Eggers asked.

"You, first."

"All right. You aced the bar exam."

"I told you I would," Carly said to Stone.

"I mean," Eggers said, "that you got every answer right."

"That's what I meant, too," she said.

"The president of the New York State Bar Association called me at the end of the workday. He said that the people who administer the exam said that this has never happened before."

"Wow," Dino muttered.

"Is that true?" Carly said.

"It's true."

"Oh, well," she said. "I guess I'll just have to learn to live with being unique in my perfection."

Stone winced. "Can't you think of some way to bring her down a notch or two?" he asked Eggers. "She's nearly impossible to live with as it is."

"Don't worry," Eggers said. "That will happen eventually, just not tonight."

"Okay," Stone said to Carly, "you get to have one free evening, then it's the real world again. What do you want?"

"Look, I was first in my class at Yale Law, I got a job at the most prestigious firm in town, I drive a snappy car, and as of tonight, I have my own apartment. What else could I possibly want?"

"Well, that was easy," Eggers said.

"Come to think of it, there is something I'd like," Carly said.

"Uh-oh," Stone said.

"If it's within your gift."

"Try me," Eggers replied.

"I'd like a larger office with a leather sofa and a TV."

"A *TV*?" Eggers said.

"A big one, that disappears into a cabinet when clients are around."

"Why do you need a TV in the office?"

"So I can learn when I'm not working. I like to keep it tuned to a news channel."

"Which one?"

"CNN or, sometimes, MSNBC, when Rachel Maddow or Lawrence O'Donnell are on."

Eggers chewed on an ice cube for a moment. "Is that it? No afterthoughts?"

"That's it, no afterthoughts."

"Okay, you'll get the office two doors down from Herb Fisher. The guy who sits there is being moved to Estate Planning next week."

"The poor bastard," Carly said. "Okay, deal!" She put out her hand and Eggers reluctantly shook it.

Dino turned to Eggers. "Next, she'll want *your* office."

"Not just yet," Carly said.

"You're going to have to make a lot of rain before you think about that," Eggers said.

"Oh," she replied, "I made a little rain today."

"What rain?"

"A liquor distributer, Jones & Jones. That's what I thought you were talking about earlier."

"Do you know someone there?" Eggers asked.

"I've been sleeping on Jones's sofa," she replied. "By the way, they spent two million on legal fees last year."

"Gee," Eggers said. "Maybe you ought to come and sit next door to me."

"All in good time," she said. "Let's frighten Herb Fisher first."

"Okay, one more thing," Eggers said. "It's traditional that when an associate passes the bar exam on the first try, we give them a ten percent raise."

"How generous!" she said.

"Because you aced it, and because you're already making rain, I'm going to give you a twenty-five percent raise."

"Thank you! I accept with gratitude."

"You just keep showing me what you've got, kiddo."

"I'll make a point of it," she said.

CHAPTER 18

Carly's suitcase turned out to be two large ones, but Fred got them into the Bentley without trouble.

"Where were you going to sleep tonight?" Stone asked, when they were on the way home.

"In a hotel, but I didn't know the name of one," she said. "That's why I brought my bags."

"How convenient," Matilda said.

"It is, isn't it? Things always seem to work out."

"I bet they do."

Fred turned into the garage, and, while Fred got Carly's luggage, Stone took her and Matilda to the house next door and showed them the vacant apartment.

"It's charming," Carly said. "How did I get so lucky?"

"I've never seen anyone impress Bill Eggers so much

or so fast," Stone said. "Come on, I'll show you the house."

He gave Carly the ten-cent tour, with Matilda tagging along, then sat them down for a cognac in the study. "Is all this going to your head?" he asked Carly.

"Of course, it is," she replied. "I'm impressed with myself."

"Honestly, I'm a bit impressed with you, too," Matilda said.

"Thank you."

"Are you going to remember everything I've told you?" Stone asked. "All my advice?"

"I have a photographic memory," she said, "and it doesn't have to be written down."

"There's a wrinkle I haven't told you about," Stone said.

"What's that?"

"Somebody is trying to kill Matilda and me, and you have to be careful not to get in the way."

"When I met you last, *two* people were trying to kill you, and we managed to handle that."

Matilda looked at Stone, surprised. "How often do people want you dead?"

"Too often."

"Well," Carly said, "your present status sounds like an improvement over last time."

"I suppose it is, in some ways, but the would-be killer employs a number of people to do his killing for him."

"Why don't you get Ed Rawls to shoot them for you? It worked in Maine."

"It may come to that," Stone said, "but we don't want a lot of people being shot in Dino's jurisdiction. He has to live with the statistics."

"Are you able to arm me?"

"I don't know. Can you restrain yourself?"

"I'm a model of self-restraint," she said.

"First, we'll have to get Dino to get you a license to carry."

"He can do that?"

"He's the police commissioner. He can hurry the process without breaking any laws. We just did so for Matilda here. I'll get you an application to fill out."

"In the meantime, I can kill with a single blow." She held up a thumb. "I took a self-defense course."

"That makes you doubly dangerous," Stone said. "Be careful what you do with that thumb while we're waiting for your license to come through."

"I'll practice thumb restraint," she said.

After seeing Carly off to her new apartment, Stone and Matilda retired to the bedroom, where Matilda immediately slipped out of her dress and into Stone's bed.

She eyed him hungrily. "Don't keep a girl waiting."

He did not, and she rewarded him with even more enthusiasm than their previous sessions.

When all was done, she said, "That was just a little reminder, so you don't forget me."

"My dear, why would I ever forget you?"

"Why, indeed?"

The following morning when Stone went down to his office, and instructed Joan on Carly's carry license application, he found a phone message from Herb Fisher waiting for him, and he called back.

"What the hell are you doing with my new associate?" Herb asked.

"She's doing it for herself," Stone replied. "You've heard the news about her bar exam score?"

"Not yet."

"One hundred percent."

"She *aced it*?"

"She did, and she's already making rain."

"That, I heard about."

"Eggers had dinner with us last night, and he gave her a twenty-five percent raise and a new office on the spot."

"Not *my* office."

"Two doors down."

"Holy shit."

"And a leather sofa and a TV that disappears into a cabinet."

"Jesus, I had to buy my own TV."

"Get used to it, pal. She's going to be a jump ahead of you every day."

"Can I give her back to Eggers?"

"He'd probably like that, but don't worry, she'll propel herself along. She won't need your help. She can make you look good, if you don't complain about her."

"Then I'll keep my mouth shut and just go along for the ride."

"That would be a smart move, if you can manage it. Oh, and don't try to get her in the sack. You'll ruin yourself around the firm, if you should manage it."

"Don't worry, I'm otherwise engaged at the moment."

"Then just hang on and enjoy watching her work her magic."

"Gotcha."

They both hung up.

CHAPTER 19

Fred drove Stone to Woodman & Weld that morning, then picked him up again right before lunch, and headed for home. They'd only gone a few blocks when he said, "I don't mean to alarm you, but we appear to have picked up a tail."

Stone didn't look back. "Anyone we know?"

"I noticed him when I was waiting for you, but that was the first time I'd seen him."

"Describe him."

Fred did so.

"That sounds like Bozo."

"The clown?"

"Yes, but not the one you're thinking of."

"If you don't mind, I'll take a few turns. To make sure I'm right."

"If it's Bozo, you are right. But go ahead."

Three turns later, Fred said, "He's still there."

"Is he being obvious?"

"No, but he's not being as cautious as he thinks he is."

Stone looked at the road ahead. "We don't want him to know we've made him. Pull over after the next light. I'll go into that building. I believe we have a couple of clients there."

Stone entered the lobby and crossed over to the first-floor café, where he ordered a coffee and then called Dino.

"Trench appears to be back at it," he said. "He's having us followed."

"By the man from the other night? Bozo?"

"One and the same."

"Do you want me to have him pulled over?"

"No. But I wouldn't mind having someone following *him*."

"In case he gets a little frisky?"

"Not the word I would have used, but yes. Call Fred. He'll give you a description of the vehicle."

"Consider it done."

Stone settled down at one of the tables, with a copy of the *Times*. After twenty minutes, he called Fred. "Is our friend still there?"

"Last I checked."

"And Dino's people?"

"They arrived ten minutes ago."

"Good. Then come pick me up."

After watching the lawyer's Bentley pull into the garage in Turtle Bay, Bozo drove on for another few minutes, stopped at the side of the road, and called Trench.

"Barrington's at his home office."

"Does he know you're watching him?"

"He doesn't have a clue."

"Huff thought that, too."

"Well, I'm not Huff."

"What about Matilda?"

"No sign of her yet."

"What's the plan?"

"I have a couple of ideas. How identifiable do you want the bodies?"

"It doesn't matter to me, as long as they're both gone."

"That's what I was hoping you'd say. What do you think about giving them a big send-off."

"How big?"

Bozo explained what he had in mind.

Trench chuckled. "I like it."

"I thought you would."

"How long until you can be ready?"

"I'll need a few hours to get everything worked out, then it'll depend on when Barrington and Matilda go out. Worse case, by tomorrow. It's not going to be cheap, though."

"How much?"

"I need to make a few calls first."

"All right. Remember, when it happens, I want to be there."

"Don't worry. I wouldn't want you to miss it."

Bozo called an acquaintance he knew named Pike Larson.

"Yes?"

"It's Bozo."

"What do you want?" Larson asked, surly as ever.

"I need to do a little shopping. Is your store open?"

"That depends on what you want."

"An accessory for a friend's Bentley."

"How big of an accessory?"

"As big as it needs to be."

"I might have what you need. It'll cost you, though."

"How much?"

"Two Gs."

"No friends-and-family discount?"

"You are neither friend nor family."

"That hurts."

"Do you want it or not?"

"I want it."

"Then come and get it. I don't do deliveries." Larson hung up.

Bozo did some quick math in his head, factoring in what else he would need and adding a pad, then called Trench back. "It'll cost four grand."

"You're not thinking of skipping out on me, are you?"

"Four grand wouldn't get me far."

"Far enough."

"The way I see it, this is my audition. Once you see how well I do, you'll find more jobs for me, and I'll make a whole lot more than four grand."

"Give me thirty minutes."

CHAPTER 20

Trench steeled himself before rapping on the door of the restaurant that his uncle sometimes used as an office. The place was located in Little Italy and closed at lunchtime.

A burly thug looked out at him from the other side of the glass, unsmiling.

"Open up," Trench said.

The thug glanced over his shoulder and said something to someone deeper in the restaurant. When he turned back, he opened the door but stood in the gap, blocking the way.

"What do you want?" the thug asked.

"I'm going to give you a break and guess that you don't know who I am. I'm here to see your boss, *my* uncle."

Instead of stepping to the side, the thug shut the door again, locked it, and disappeared.

"Hey, asshole!" Trench knocked on the glass. "Let me in."

He raised his palm to slap the entrance again when a new man appeared. This one Trench knew. He was a member of his uncle's inner circle, Gregor . . . something or other. It didn't matter. Everyone called him the Bean Counter.

The man opened the door. "Hello, Trench. I'm sorry, but your uncle's in the middle of a meeting."

"That's okay. I'm sure you can help me instead."

"What can I do for you?"

Trench looked left and right down the sidewalk. "Can we talk about this inside?"

The Bean Counter waited a beat before he stepped out of the way and let Trench in.

There were a dozen men in the restaurant, most of them enforcer-types. The door to the private dining room behind the bar was closed, so maybe his uncle really was in a meeting. Trench didn't really care. He'd rather not deal directly with his uncle about this anyway.

"So?" the Bean Counter asked.

"I need money."

The man did not look surprised. "How much?"

"Ten grand." Trench subscribed to the belief that it was always better to ask for more than he needed.

"Ten? What do you need ten thousand dollars for?"

"Expenses."

"What kind of expenses?"

"*Personal* expenses. Is that a problem?"

"Do you think your uncle has ten thousand in cash just lying around?"

Trench knew he did. He'd seen that much and more on other visits. "Well, maybe I should ask him."

He walked around the man and headed toward the office door.

"Go ahead," the Bean Counter said. "But you should know, he's not in a good mood."

That stopped Trench. Dealing with his uncle in the best of circumstances was nerve-racking enough, but if the man was in a bad mood, seeing him was the last thing Trench wanted to do.

"I can give you five thousand," the Bean Counter said.

"That's half of what I asked for."

"Your math skills are as sharp as ever. Try this calculation out. I could give you half, or, if you keep complaining, I could give you nothing."

Five grand was more than enough, but Trench didn't want the son of a bitch to know that. Looking annoyed, he said, "If that's all you can do, I guess I'll make that work."

"I thought you might. Give me a moment."

The man vanished into a back hallway and returned holding a thick envelope. He handed it to Trench.

"Five thousand in fifties. I hope that works for you."

"Yeah. Sure. It all spends the same." Trench shoved the envelope into the inside pocket of his jacket and turned to leave.

"Trench?" the Bean Counter called.

"Yeah?"

"So that there's no misunderstanding, we'll be deducting that from next month's allowance."

Trench's cheeks flushed. This guy was seriously pissing him off. But he knew it would be a waste of energy to argue the point. Better to talk to his mother and have her take it up with his uncle later.

"Whatever."

He made a beeline for the exit.

Once he was back in his car, he called Bozo.

"I've got your money."

"Great. Where can I meet you?"

"The gym?"

"I'll be there in fifteen."

CHAPTER 21

Joan walked into Stone's office with the *New York Post.* "Have you seen the legal news on Page Six?"

"That's the gossip page, not the legal news."

"It is today," she said, handing him the paper, folded to the headline.

ASPIRING ATTORNEY ACES BAR EXAM!!!

Never been done before. The examiners won't say who, until the grades are officially announced, but he's at one of the top firms.

"They say it's a guy, see?" Joan giggled. "I can't wait until they have to correct that."

"How do you know about this, Joan?"

"I know about everything, didn't you know?"

"You've been eavesdropping, then?"

"You've all been shouting it from the rafters. Why do I need to eavesdrop?"

"Oh."

"I hope they gave her a raise," Joan said.

"Twenty-five percent, and you may as well know it all. She got a bigger office, a sofa, and a TV that disappears into a cabinet when a client shows up. Now, keep it to yourself, will you?"

"This is the one who's moved into the flat next to mine?"

"Have you moved back in already? If so, this is the first-known case of somebody moving out of a twenty-room house and into a four-room flat."

"I couldn't sleep in that place for another night. It just wasn't me!"

"Well, before you know it, she'll be moving into a twenty-room house! She's that smart."

"I guess you won't need me anymore, huh?"

"Not after you've briefed her on the security system codes and instructions, but stay on, anyway. I'm used to you."

"Gee, thanks!" Joan flounced out.

"And don't flounce!"

Joan slammed the door, for emphasis.

Later, Matilda entered Stone's office. "Knock, knock."

"Most people knock *before* they come in."

"I find it more interesting to keep people on their toes. Are you still busy?"

"No more than usual."

"Can I assume you are planning on eating dinner?"

Stone glanced at his watch. It was nearly six PM. He'd spent most of the afternoon going from one client phone conference to another and had lost track of time. "If there's a meal I never miss, it is that."

"Oh, good. I was afraid I would have to go it solo."

"That would be a crime. Patroon? Leave in thirty minutes?"

"It's like you're reading my mind. I'll get changed."

"And I'll call Dino."

As Stone and Matilda were getting into the Bentley, Carly pulled into the garage.

"Hungry?" Stone called to her.

"Famished."

"Then you should join us."

The look Matilda gave Stone as he climbed in beside her wasn't exactly annoyed, but it wasn't not annoyed, either.

"My dear, your green is showing."

"I have no idea what you're talking about."

Carly popped her head through the doorway a moment later. "I hope I'm not imposing."

"Not at all," Matilda said, instantly cheery.

"We're meeting Dino," Stone said. "So, you can be his date."

"No Viv?"

"Working, I'm afraid."

"My luck, then."

"Please don't tell him that, I'll never hear the end of it."

They were halfway to Patroon when Fred said, "Our friend is back."

"What friend?" Matilda asked.

Stone told her and Carly about being followed that morning. Matilda started to turn her head so she could look out the back window.

"Don't," Stone said. "We don't want him to know we know he's there."

"Do you know who it is?" Carly asked.

"I personally haven't seen him, but from Fred's description, it's Bozo."

"The clown?"

"You and Fred should hang out more. This Bozo works with our friend Trench."

Matilda huffed. "Trench is not my friend."

"Nor mine." Stone called Mike Freeman, since Dino would also be on the way to Patroon. "I have a situation I could use some help with." He explained about the tail.

"How do you want us to handle it?"

"Keep an eye on him and be ready in case he tries anything."

"How likely do you think that will be?"

"Highly."

"Understood. Where are you headed now?"

"Patroon."

"I should have guessed. Have Fred take the long way, and I'll have a car following within ten minutes. I'll message you when you can go eat."

"Don't take too long. I have two ladies here who might start gnawing on my arm, if you do."

"That sounds like the kind of problem you enjoy."

"I'd rather not test the theory."

Dino was waiting for them at the bar when they arrived.

Carly took in Patroon with a sweeping glance. "Oh, my!" she said.

"Didn't New Haven's finest rank with this?" Stone asked her.

"I dismissed New Haven's finest the first month of my freshman year," she said. "This place has New York written all over it."

"How discerning you are," Stone replied, grinning.

"Oh, I am. I am."

"I thought you said seven," Dino said.

"Sorry," Stone said. "We had to take a slight detour. But we're here now."

"And we've brought you a date," Carly said. "Me."

"Please, don't tell Viv that," Dino said.

"Why does everyone keep asking me not to tell people things?"

CHAPTER 22

"That's them." Bozo pulled to the curb around the corner from Patroon, where two people stood waiting. The man, Reggie Hogan, was another trainer at the gym, and the woman, Candy Parker, was a waitress at a diner Bozo frequented.

Trench eyed the couple. "Are you sure you can trust them?"

"Stop worrying. They'll be fine."

"If you say so."

"I do. Now, wait here. I'll be right back." Bozo hopped out of the car and approached the couple.

"You two ready?"

"Aren't you forgetting something?" Hogan held out his hand.

"You get paid *after* you do the job."

"How do we know you won't stiff us?"

Bozo moved into Hogan's personal space. "What did you say?" They were the same height, but both knew Bozo would wipe the floor with Hogan if blows were exchanged.

"Nothing. Sorry."

"That's what I thought." Bozo stepped back. "I ask again: Are you ready?"

The woman nodded. "I am."

"Me, too," Hogan said.

"You need to keep the driver busy for at least ten minutes," Bozo said to Candy. "Not a second less."

"I remember."

"Do you also remember what to do after?"

"I find a seat at the bar and keep an eye on the attorney and his friends. When they're leaving, I let you know."

"Perfect. Okay, let's get this show on the road. Make it look good."

Fred knew something was up the moment he saw the man and woman heading down the sidewalk. They appeared to be arguing, but it was more of an act than the real thing.

Fred donned his wireless earbuds, and as instructed, called the number Stone had passed on to him.

"This is Fred Flicker. To whom am I speaking?"

"Janet Hutto, Strategic Services."

"Ms. Hutto, can you see me from your position?"

"We can."

"And the couple walking toward me?"

"We see them. They look like they're acting."

"That is my assessment as well."

When the couple was almost to the Bentley, the man slapped the woman, pushed her to the ground, and ran off. She cried out, then called for help.

"She's baiting you," Hutto said.

"Perhaps I should find out why."

"Would you like me and my partner to back you up?"

"Don't. That might scare her off. I can handle myself."

"If you're sure."

"I am."

Fred slipped his phone into his pocket, leaving it on, then climbed out and jogged to the woman.

"Madam, are you okay?"

"Far from it. I was just assaulted."

"I saw. Can you stand?"

"I think so."

Fred helped her to her feet.

"Would you like me to call the police?"

"I'm—I'm not sure. Is there someplace I can sit down and catch my breath?" She looked around, until her eyes landed on the entrance to Patroon. "Maybe in there?"

"I'm sure they have a chair you can use."

She took a step and winced. "Can you help me? It feels like I twisted my ankle."

"I would be happy to."

In his ear, Hutto said, "We'll keep an eye on your car."

Fred escorted the woman into the restaurant and over to a chair, then approached the hostess. "Someone pushed her down. Is it possible to get her a glass of water?"

"Of course."

As the hostess hurried off, Fred looked back at the woman. "Have you decided if you'd like me to call the police?"

"I don't know. I wouldn't want to cause a scene."

"I think it's more accurate to say that man who hit you and shoved you onto the sidewalk caused the scene."

"I need a few minutes to collect myself before I decide anything."

"Very well. If you need me, I'll be out at my—"

"No," she said, grabbing his hand. "Please, stay with me. I'm afraid he might come back."

"Mr. Flicker," Hutto said over Fred's earpiece. "A man has just stopped at the back of the Bentley and dropped down, out of sight."

Fred took the seat next to the woman. "Of course, I'll stay."

Stone's phone buzzed in his pocket. He pulled it out, saw it was a call from Mike Freeman. "Hello, Mike."

"There's a situation with your car," Mike Freeman said.

Stone covered his phone, and said to his companions, "I'll be right back." He found a quiet spot near the restrooms and brought his phone back up. "What kind of situation?"

Mike told him about Fred and the woman, then said, "After they went inside, a man approached your car and crawled underneath it. I've been told he just left."

"Describe the man."

Mike did.

"That sounds like Bozo. And before you say anything, no, not the clown."

"I'm going to send one of my people to check his handiwork."

"Good idea. Let me know what they find."

"I will."

Stone returned to his friends.

CHAPTER 23

Stone's dinner of pan-seared halibut had just been set in front of him when Mike called him again.

"It appears Bozo has ill will toward you. He has attached a car bomb to your Bentley."

Stone excused himself from the table again. "Can it be removed?"

"I am told it can."

"Without any harm to your agent?"

"I don't think she would have suggested it if that was a possibility."

"I suppose not."

"Would you like her to remove it?"

"I would," Stone said. "Very much."

"I'll let her know. We'll get it someplace safe and—"

"Actually, Mike, I have something else in mind."

He told Mike what he was thinking, then, after they

hung up, he messaged Fred with some instructions before heading back to eat.

An automated voice came through the earbud in Janet Hutto's ear, announcing she had a call from Mike Freeman. She verbally accepted and Mike was put through.

"Are you still near the bomb?" he asked.

"It's right above me," she said, from her position lying underneath the Bentley.

"Excuse me for double-checking, but you are sure you can remove it without accidentally setting it off?"

She'd gone over the specifics before, but she appreciated his caution. "Yes, I've stabilized it and can safely transport, so it should be a snap."

"Then do it. I'll stay on the line."

Hutto cut the tape holding the bomb in place and scooted out from under the Bentley with the device cradled in her arm.

"I'm clear. What should I do with it?"

"Fred should have remotely unlocked the trunk while you were under the car."

She tested the trunk and it moved upward. "Yes. It's open."

"Put the device in there and return to your vehicle."

Hutto did as instructed.

"How are you feeling now?" Fred asked the woman.

"A little better, thank you."

"Any further thought about phoning the police?"

She shook her head. "I'd rather not. I don't need that kind of trouble."

"Very well. If you're all right on your own, I should get back to my car."

She sneaked a look at her watch, but Fred didn't miss it. Apparently, the amount of time she'd been told to keep him busy had passed, because she said, "I'm fine. Though I could use a drink."

"May I suggest the bar? I understand they have a wide selection of whatever you might want."

"What a lovely idea. Care to join me?"

"Not while on duty, I'm afraid."

"That's a shame." They stood. "Thank you for your help."

"Of course."

The woman headed into the bar and Fred returned to the Bentley.

There, he retrieved the bomb from the trunk and then went for a walk.

CHAPTER 24

"I don't think I've ever had a bad meal when I'm with you," Carly said.

Matilda eyed her. "Have you eaten a lot of meals with Stone?"

"That depends on your definition of 'a lot.'"

"More than three?"

"That seems like a low bar, but, yes, more than three."

Matilda turned to Stone, pressing her softest spots against him. "Do you always treat your junior attorneys so well?"

"Carly is no ordinary junior attorney."

"Oh, that's right. The whole acing-the-bar-exam thing."

"It's more than a thing," Stone said.

"Of course. My apologies. I didn't mean to trivialize it." She gave Carly a weak smile. "Smart and beautiful. Congratulations again."

"Thank you."

Stone's phone rang. Fred.

"All done here. Ready to go when you are."

"Thank you, Fred. Were you able to get a positive ID on the driver?"

"It's Bozo."

Not a surprise, but it was nice to have it confirmed.

"And he's not alone," Fred added.

"Oh?" This *was* a surprise.

"Trench Molder is with him."

"Is he now? Not his lucky day, I guess."

"No, it is not."

"His bad luck is our good. Thanks again, Fred. We'll be out in twenty." Stone hung up.

"Are you going to tell us what's going on now?" Matilda asked.

"All in good time."

"I'm curious, too," Dino said.

Stone gave him a pointed look. "Don't be."

"Understood," Dino said, then looked around the table. "A glass of port before we leave?"

"It would be uncivilized not to," Stone said.

Bozo's phone rang. "Yes?"

"It's Candy. They're heading out."

"Thanks. Come by the gym tomorrow morning and I'll pay you."

"Will do."

He started the car.

"We're on?" Trench asked.

"We're on."

Trench sneered. "Finally."

"Fred?" Stone asked.

"Still there. About six car lengths back."

Bozo had begun following them again as soon as they'd driven off from Patroon.

"Are we far enough away?"

"We should be."

"I was hoping for something a bit more definitive."

"Probably?"

"We need to get you a better thesaurus."

"I'll pick up one tomorrow."

"What are you waiting for?" Trench asked.

They'd been following the Bentley for ten minutes, and Trench was running out of patience. Each tick of the clock was an additional second of life neither Barrington nor Matilda deserved.

"For traffic to thin a bit. Unless you don't care about collateral damage."

In the grand scheme of things, Trench didn't, but

additional casualties would take away focus from the two he wanted dead. "Fine. But don't wait too long."

The opportunity came two minutes later, when traffic spread out enough to create a clear zone between them and the Bentley.

Bozo lifted the remote that would trigger the car bomb. "Ready?"

"Let me do it."

Bozo had been looking forward to pressing the button, but as Trench had pointed out before, Bozo wasn't the one paying the bills. He handed it over.

Trench aimed the device toward the Bentley. "Good riddance, assholes."

He pressed the button.

The street lit up with an explosion that ripped Bozo's car apart. The shock wave rocked the Bentley, but Fred kept control and presently had them speeding away.

"What was that?" Matilda asked.

She, Stone, and Carly all looked out the back window.

Lying in the middle of the street, and growing more distant by the second, was the flaming hulk of Bozo's sedan.

"That is the end of your problems with Trench Molder," Stone said.

"What do you mean?" Her eyes widened. "Are you saying he was in that car?"

"He and his buddy Bozo."

"You blew him up?"

"I did no such thing. He blew himself up."

Matilda looked no less confused.

"While we were having diner, Bozo put a bomb under the Bentley. Fred simply returned it to its rightful owners."

"So, that should have been our car burning back there," Carly said.

"That was their plan."

The blood drained from Matilda's face. "Oh, my God. I . . . I knew he wanted us dead, but I didn't think he'd actually try to do it."

"Try, yes. Succeed, no." Stone pulled out his phone and called Dino.

"Miss me already?"

"I thought you'd like to know you're about to receive word concerning an exploding car."

"Not yours, I'm guessing."

"Very astute. I'm not sure who the vehicle belongs to, but I can tell you who was inside."

"I'm all ears."

"Trench Molder and his pal Bozo."

"Next you're going to tell me you had nothing to do with it."

"It's like you're reading my mind."

"Anyone hurt on your end?"

"Not a hair on a head."

"Is there more I should know?"

"Officially, no. Unofficially, the bomb was meant for us."

"Then unofficially, I'm glad it failed."

"That makes five of us."

"That's my other line ringing. Gotta go."

"Act surprised."

"It won't be the first time."

CHAPTER 25

Matilda held out her empty glass. "Another, please."

Stone poured her another two fingers of Knob Creek. "If you sip it, it will last longer."

"Tell me that on a day when someone doesn't try to blow me up." She downed the whiskey and held out her glass again.

This time, Stone poured a single finger, which she disposed of in the same manner, then stood. "If you'll excuse me, I think I'll go lie down."

Without another word, she headed up to the master bedroom, leaving Stone and Carly in the study.

"I don't think she's taking Trench's death well," Carly said.

"It's not so much *his* death, but that it could have been hers."

"Clearly, she hasn't hung around you long enough, or she would be used to it by now."

"Are you saying you are used to it?"

"Funny, even when we had that trouble in Maine, the threat of death didn't bother me. In fact, it was kind of exciting."

"I'm not sure that's the right attitude to have."

Carly shrugged. "I can't help my attitude. It just is."

"I suppose, but do yourself a favor, and don't mention that to anyone else. They might not be as understanding."

"Again, with the warnings."

"Believe me, it's for your own good."

"I guess I'll have to take your word for it."

"You do."

Dino entered the study. "I'm glad to see you both in one piece."

"Technically, we are two pieces," Carly said.

"Technically, I'm still glad."

"Any news yet?" Stone asked.

"The car that's not a car anymore was registered to one Thomas Bozeman, known to his friends as Bozo. I can confirm there was one other clown with him."

"Trench Molder."

"That won't be official until the DNA test is performed. But I can tell you that a singed and partially intact wallet was found containing ID for said individual."

"Drink?" Stone asked.

"I thought you'd never ask."

Once Dino had a glass in his hand, he said, "Where's our Matilda?"

"She's called it an early night," Stone said. "I'm afraid the bombing has rattled her."

"Something like that would rattle anyone."

"Not me," Carly said.

"Carly, what did I tell you?" Stone said.

"Oh, right. Sorry." She clamped a hand over her mouth.

"It didn't scare you?" Dino asked her.

She glanced at Stone, then shook her head.

"You should probably not tell anyone that."

"So I've been told."

CHAPTER 26

"Wait here," Gregor Dryga, aka the Bean Counter, told his driver. "I won't be long."

He entered the building and went to the private elevator that serviced the penthouse. As usual, two men guarded the entrance.

"Is he in?" the Bean Counter asked.

"He is," one of the men said, "but he left instructions not to be disturbed."

The Bean Counter knew this meant his boss was not alone. "This is a priority one situation."

With those magic words, the men stepped to the side, and the Bean Counter entered the elevator.

Alexei Gromyko—who, like his late brother, was also known as the Greek—pulled loose the towel that had been wrapped around . . . Brenda? Brianna? He couldn't remember her name, but it wasn't important. After tonight, he would never see her again. At the moment, however, he was seeing everything, and it pleased him.

"That's not fair," she said. "You still have your towel on."

"Then perhaps you can do something about it."

She smiled and dropped to her knees. As she began pulling his towel away, someone knocked on the door to the master suite.

"Are we expecting company?" she asked.

"We are not."

A second knock was followed by, "Sir?"

Gromyko grimaced and barked, "What?"

"Mr. Dryga is here. He says he needs to speak with you."

The woman gave Gromyko an invigorating rub.

"I'm in the middle of something," he called out.

"I'm sorry, but he said it was 'priority one.'"

"Shit." Gromyko took hold of the woman's hands and lifted them away from his body.

She pouted.

"Watch TV or something. This shouldn't take long."

He rewrapped the towel around his waist and walked into the living room where the Bean Counter waited with the guard who'd knocked on the door.

"This better be good," Gromyko said.

"Good is not how I would describe it," the Bean Counter said and shot a glance at the guard.

Gromyko nodded at the man. "Leave us." Once the bodyguard was gone, the Greek asked, "What is it?"

"Your nephew is dead."

Gromyko stared at him, not quite understanding the words. "Say that again."

The Bean Counter did.

"You mean Trench?"

"I'm afraid so. It's not confirmed yet, but I received a call from one of our people in the police department, and he said it's only a formality they are waiting on."

Gromyko walked over to the bar, poured himself a whiskey, and drank it in a single shot. "What happened?"

The Bean Counter described what he knew about the incident.

"A car bomb? So, this wasn't an accident. Someone killed him on purpose."

The Bean Counter was tempted to say that's how car bombs worked, but he knew that would not be received well. Instead, he went with, "So it appears."

Gromyko did not like his nephew. Trench was a lazy screwup who spent his life bouncing in and out of trouble that Gromyko had to clean up. The Greek had tried to give him a job in the organization, but Trench had half-assed it so badly that Gromyko had realized it was best to keep the waste-of-life far from the business.

Trench's saving grace was that he was Gromyko's sister's only son. The Greek loved her more than he loved anyone else, so as much as he would have liked to disown Trench, that wasn't in the cards. And now he would have to call his sister to tell her the news.

"Find out who did this," he said. "Then I will destroy him."

"When I know, you will know."

As the Bean Counter left, Gromyko stormed back into the bedroom. The woman was in the bed, the sheet barely covering her.

"I was starting to think you forgot about me."

"Get out."

"What?"

"Get out now."

He marched to his dresser, pulled out several one-hundred-dollar bills, and tossed them at her. "For your time."

"Hey, I'm not—"

The stare he gave her cut her off. She gathered the money, dressed quickly, and was out of there in a flash.

CHAPTER 27

Stone woke the next morning to find the bed beside him empty and his secure cell phone ringing. From the lack of a caller ID, he knew who it was.

"This is early even for you, Lance," he answered.

"It's not every day one of my consultants is almost blown up."

"Your concern is touching."

"It's not you I'm concerned about. It's the fact that if you died, I'd have to find a replacement."

"Again, I'm touched. Is there another reason for your call other than to remind me how important I am to you?"

"There is. I'm afraid you might have kicked a hornets' nest."

"How so?"

"If my information is correct, last night's car bomb

had been intended for you. But it was relocated before it was set off."

"Let's pretend that's correct. What does that have to do with a hornets' nest?"

"Do you know who Trench Molder is?"

"I believe you mean 'was.'"

"The question is the same."

"He was a self-important layabout with too much time on his hands."

"He may have been that, but he was also related to the Greek."

That stopped Stone. He'd had an up-close and personal relationship with Serge Gromyko, aka the Greek. The man was the former head of the Russian mob. Former because Stone had ended the Greek's life with a bullet to the head.

"Trench was a Gromyko?"

"No," Lance said, "but his uncle is."

"Again, I think you are confusing your verb tenses. The Greek is dead."

"I'm talking about the new Greek."

"The new Greek?"

"Alexei Gromyko."

"There's another Gromyko?"

"Serge's half brother. Took over after you removed Serge from the picture. And before you ask, he's also half Russian, half Greek. Their mothers are sisters."

"That's a disturbing family tree."

"The important branch to you is that Alexei's sister is Trench's mother."

"And now you're going to tell me this new Greek and Trench were close."

"Close or not, he will not be pleased his nephew is dead."

"The bomb wasn't mine, and I didn't set it off. Trench killed himself."

"Do you think that's how the Greek will see it?"

"Gee, thanks for being a bright ray of sunshine."

"I am only the friendly messenger. My suggestion is to take the appropriate precautions."

The line went dead.

Stone sat up, his mind churning through this new information. It took him several moments before he realized Matilda was standing in the room, fully dressed, and with her travel case beside her.

"Going home for more clothes?" he asked.

"I'm going to visit my sister."

"Will you be back in time for dinner? Or . . ."

"She lives in Los Angeles."

"So that's a no, then."

She expelled a breath, then sat on the bed near him. "Thank you for letting me stay here with you. On the whole, it's been a wonderful time."

"I sense a *but* coming."

A weak smile. "But what happened last night seems to have knocked me off-kilter."

"Explosions have a way of doing that."

"My sister has been asking me to visit for a while, and now seems as good a time as any. Better, really."

"How long will you be gone?"

"I don't know. Are you going to try to talk me into staying?"

Given the turn of events, Stone thought her leaving town was an excellent idea. But there was no need to increase her anxiety. "Would it work if I did?"

"Maybe, I don't know." She thought for a moment, then shook her head. "No. I need to clear my mind, and I can't do that here in the city."

"I understand."

"Maybe I'll do some painting. Who knows?" She leaned over and kissed him tenderly. "I'll miss you. But something tells me you'll find new companionship soon enough." She smirked knowingly, then stood. "Well, I'll be off."

"Surely, you can have breakfast first."

"No time, I'm afraid. I've already purchased my ticket and need to get to the airport. All I have to do is call a cab."

"Nonsense. Fred can take you. Let me get dressed, and we'll get you sorted."

Ten minutes later they were in the garage, with Matilda's bag in the trunk of the Bentley.

"Thank you, again. You saved my life," she said and gave him a hug.

"We'll see each other soon."

"Maybe."

As she pulled away, the sound of someone entering the garage caused them both to look over.

"Hi," Carly said. She was dressed for a day at the office. "What's going on here?"

"Matilda's off to California to visit her sister," Stone said.

"Oh. Have a good trip."

"I plan to," Matilda said. She started to climb into the Bentley, then stopped and glanced back at Carly. "He's all yours now."

Before Stone could react, she entered the car and closed the door behind her.

He shrugged and said to Carly, "I have no idea what that was all about."

"That's okay. I do." She got into her car and drove out right behind the Bentley.

CHAPTER 28

Across town, the Bean Counter's secretary, Lauren, stuck her head into his office. "He's here."

"Show him in."

Moments later, his protégé, Leonid Korolev, entered. A six-foot-four slab of muscle wrapped in a three-thousand-dollar Armani suit, Korolev was an intimidating sight. Worse, at least to those who crossed the family, he was both smart and clever, two things that did not always go together.

"You wished to see me, sir?" he said.

The Bean Counter motioned to the chair on the other side of his desk. "Have a seat." When he did, the Bean Counter went on, "Are you familiar with Trench Molder?"

"Greek's nephew? I have met him once or twice." There was no missing the disdain in Korolev's voice.

"You won't have to worry about running into him anymore."

"Sir?"

"He's dead."

Korolev's only reaction was a slight raise in his eyebrow.

"But as I'm sure you can guess, the Greek is not happy."

"How did Trench die?"

"Car bomb."

This time Korolev did not even attempt to hide his surprise.

"I need you to find out who is responsible," the Bean Counter said. He set a piece of paper in front of his lieutenant that had three phone numbers on it. "Memorize these then destroy this. They are the numbers of friends of the family who work on the police force." In other words, men on the payroll who could tell Korolev what the police knew. The Bean Counter laid a key on top of the paper. "And this is to Trench's apartment."

Korolev picked the items up. "Do we think it was a rival organization?"

"At this point, I know as much as you do. It's your job to figure it out." The Bean Counter hoped it wasn't another family. That kind of conflict was bad for business. "And make sure you get it right. The last thing we want is to go after the wrong people."

"I understand."

"If you need any other resources, let Lauren know."

Korolev took that for the dismissal it was and headed for the door.

"Leonid," the Bean Counter said.

Korolev turned back.

"The sooner you have the answer, the better. By tomorrow would be best. The day after at the latest. Beyond that . . ."

The look in his eyes told Korolev exactly what would happen in that instance.

Korolev called the first number from the list.

"Samuels," a man answer.

"Officer Samuels, the Greek sends his greetings."

The sound of muffled movement was followed by Samuels whispering, "This really isn't a great time."

"I'll be sure to let the Greek know that."

The cop cursed under his breath. "Okay, okay. Give me a second." It was more like twenty before he came back on and said, "How can I help you?"

"Email me everything the police have on the car bombing from last night." He recited the throwaway address he'd created for this purpose.

"I, um—"

"You're not going to tell me you can't do it, are you?"

"No. No. I—I'll get it. I just need a little time."

"You have one hour. If I do not receive it by then, I'll assume you are refusing to cooperate."

"Hey. There's no reason to—"

Korolev hung up.

Forty minutes later, Samuel's email arrived, with an up-to-the-moment police report on the incident attached.

Unfortunately, the police did not have much about the bombing so far.

The IDs of the victims—there were two—were still pending DNA tests. Trench's name was listed, along with that of Thomas Bozeman, the owner of the car.

Bozeman worked at the gym that Trench used, so that's where Korolev headed.

One of the first things he learned when he arrived was that Bozeman, who apparently went by Bozo, was the second gym manager to die that month. Both due to unnatural circumstances.

He also found out a trainer named Reggie Hogan had done a job for Trench and Bozo only an hour or so before the bomb went off. Hogan knew little, though, and was more hung up on the fact that he hadn't been paid for the work. He did reveal that there was a woman on the job with him, who had played a larger role. For a hundred bucks, the man gave him the woman's name and the location where she could be found.

Candy Parker started work at the diner at four PM. Korolev arrived at 4:15 and took a seat at the counter.

After she poured him a cup of coffee, he said, "You're Candy Parker, correct?"

"I am. Do we know each other?"

"We have a mutual acquaintance. Trench Molder."

"Trench? That's Bozo's friend, right? I've seen him but never met him."

"I'm sorry to be the one to tell you this, but they were both killed last night?"

"What? Oh, my God!"

The outburst drew the attention of one of the other waitresses. "Candy? Everything all right?"

"Watch the counter," Candy said, her eyes filling with tears. "I'll be right back." Without waiting for a response, she disappeared through a rear door.

Korolev followed and found her in the alley behind the restaurant, sitting on her heels, her back against the wall.

"I'm sorry," he said. "I know it's shocking."

It took a moment before she looked up at him. "How?"

"Someone blew up the car they were in."

She put a hand over her mouth. "I went into the gym this morning to see Bozo, but was told he hadn't arrived yet. I just saw him last night."

"Near Patroon."

"Yes. How did you know?"

"I talked to Reggie. He told me why you were there. He said you might know the names of the people Trench and Bozo were interested in."

She shook her head. "No one told me their names. Bozo just gave me a description of the man he wanted me to watch. He called him 'the attorney.'"

"Can you give *me* his description?" That would be better than nothing.

"Are you a detective?"

"Trench was related to a . . . friend. I'm helping find out what happened."

"Oh, I'm so sorry."

"The description?"

"Of the man? Or do you want descriptions of the women who were with him, too?"

"All of them, please."

When Korolev arrived at his apartment that evening, a list of green Bentley owners in the city was waiting for him. This was courtesy of a woman who worked at the Department of Motor Vehicles, another "friend" of the family.

The list wasn't long and included driver's license photos of as many of the owners as possible. Korolev initially had high hopes of finding a match to one of Candy's descriptions among them, but it was not to be. Most of the vehicles were owned by corporations, and none of the private owners resembled the people Candy had seen.

Korolev called the Bean Counter's office.

"Hello, Leon," Lauren answered. "What can I do for you?"

She was the only one in the organization that called him by his childhood nickname, something he usually hated, but not when she did it. "How is my favorite executive assistant?"

"Tired. Hungry. Dying for a drink. But I assume that's not why you called."

"It could be," he teased.

"Leon, you know I don't like it when you tease me like that," she said, clearly liking it. "What can I do for you?"

He told her what he needed.

"That should be possible," she said. "Can you hold, and I'll check?"

"For you, anything."

"If only that were true."

It was more than three minutes before she came back on the line and said, "I'm told you will have the information by ten AM tomorrow, latest."

"Come on, Lauren. You know my deadline. You can't do better than that?"

"If I could, I would. But what I believe you wanted to say was 'Thank you, Lauren. I appreciate your help.'"

"Thank you, Lauren. I appreciate your help."

"As you should. But if you really want to thank me, you can buy me a drink."

"I would love to show my gratitude by buying you a drink."

"What a lovely idea. Meet me in an hour. Nubeluz at the Ritz. I assume you know it."

"I know it."

"Don't be late."

CHAPTER 29

Earlier that day, Joan walked into Stone's office. "It's time."

"Time for what?"

"Your follow-up doctor appointment."

"Follow-up for what?"

She picked the blackjack up off his desk and mimed hitting herself in the back of the head with it. "Ring any bells."

"If memory serves, it rang a lot of bells," he said. "I don't need to go to the doctor. I feel fine."

"And we want you to stay that way, don't we? Now, come on." She moved around his desk and shooed at him until he got up. "Fred's waiting for you in the garage."

"I'm not going to be able to talk you out of this, am I?"

"What do you think?"

"Fine. I'm going. I'm going."

A few minutes later, he was riding in the back of the Bentley.

"Fred, we haven't picked up any new friends, have we?"

Fred glanced at Stone, in the rearview mirror. "I'm sorry?"

"I mean, like last night."

"Oh." Fred used his mirrors to check the road behind them. "No. We are quite alone. Are you expecting more attention?"

"I'm afraid that may be a possibility."

"Then I will keep an eagle eye out."

"Thank you, Fred."

Stone arrived at the hospital right on time and was soon joined in an examination room by Dr. Herbert.

"You look a lot better than you did the last time I saw you," Dr. Herbert said.

"I feel a lot better than last time."

"Any more headaches?"

"Not for a few days."

"Dizziness? Vertigo?"

"Same."

Dr. Herbert checked the back of Stone's head, then placed his stethoscope against Stone's chest. When he finished, he said, "And you haven't hit your head on anything?"

"Just my pillow when I go to bed."

"Keep it that way, and you should be fine." He wrote something on a tablet computer, then said, "I'd like to see you again in another month."

"Is that really necessary?"

"Probably not, but best not to take chances, don't you think?"

On the way back home, Dino called.

"I got the results from the rapid DNA tests. The bodies officially belonged to Trench and Bozo."

"Did Lance call you?"

"No, why?"

"Trench was the Greek's nephew."

"The old Greek or the new Greek?"

"How do you know about the new one?"

"Word gets around fast in my circles."

"Well, it's the new."

"Shit."

"My thoughts exactly. Come over for dinner tonight?"

"Can I bring Viv?"

"I would be insulted if you didn't."

Dino and Viv arrived at six-thirty for drinks before dinner. Stone had just poured Dino his second Knob Creek when Carly called. "I'm sitting alone in my new apartment and thought you could use some company."

"I have company," Stone said, "but you are more than welcome to join us."

"Oh, I didn't realize. I don't want to bother you."

"It's Dino and Viv."

"I'll be right over."

When he hung up, Dino asked, "Who will be right over?"

"Carly."

Viv smirked at Dino. "Your date from last night."

Stone held up his hands. "I didn't tell her."

"No, you didn't," Viv confirmed. "Dino outed himself."

"I don't know what came over me," Dino said.

"I can verify he kept a respectable distance from Carly at all times."

"I have no doubt." Viv patted her husband on the cheek. "Such a good husband."

Stone let Helene know there would be one more for dinner.

Carly arrived moments later. She surveyed the room, then looked at Stone. "So, tonight I'm *your* date."

"See," Dino said. "This is why I said something to Viv."

"Said something?" Carly said. "What did you say?"

"Trust me," Viv said. "It's not important." She patted the sofa beside her. "Come. Sit by me."

Stone poured Carly a drink and retook his seat.

"Since you didn't call again this afternoon," Dino said to Stone, "I take it you didn't have any trouble."

"Trouble?" Carly said. "I thought you got rid of your trouble last night. Don't tell me someone *else* is trying to kill you already."

"That remains to be seen," Dino said.

Stone told her about Lance's revelation concerning Trench's family connections.

"Surely, you can convince them this wasn't your fault. Trench was trying to kill you. Me and Matilda, too, for that matter."

"I don't think they'll care."

"Stone and the Russians have a history," Dino said.

Carly frowned at Stone. "You really need to get better at making friends. It's causing me to question all the advice you've given me."

"My advice is sound," Stone said. "Those who aren't my friends, aren't my friends for a reason."

"He's not wrong," Dino said.

"Oh, all right. I'll continue to listen to you."

"Stone," Viv said, "I think having some of our people watch over you for the next few days wouldn't be a bad idea."

"Better hers than mine," Dino said. "If cops are hanging around you, it'll be easier to tie you to the bombing."

"As much as I'd like to say it's not necessary," Stone said, "I think you're right, Viv, and I would appreciate it."

Viv pulled out her phone and made a call to Strategic Services. When she finished her conversation, she said, "There will be a rotating team watching your place until you think it's okay to call them off."

"Thank you."

Helene entered the room. "Dinner is ready."

CHAPTER 30

Korolev woke with his arm draped over Lauren.

Last night, one drink had led to two, two to three, and three to a cab ride back to her place.

Lauren moaned, then stretched her arm and turned so she was facing him. "Good morning."

"Good morning, yourself."

She looked at him, a mischievous smile on her lips. "I thought you might disappear during the night."

"Not a chance."

She laughed. "Oh, you *are* different. I thought you might be."

"What? Did you think I was a love-'em-and-leave-'em type?"

"One can never know."

"Well, I'm not."

"I guess I'll have to test you on that."

"And what does your test entail?"

Her hand slipped between his legs, and she showed him exactly what that entailed.

Thirty minutes later, they moved to the shower where they gave each other a thorough cleansing.

Back in the bedroom, Korolev pulled on his pants, and then checked his phone for messages. He was pleased when he saw an email from Lauren's contact, with the information he'd requested. "Your contact came through."

Lauren straightened the dress she'd just put on and looked at her watch. "Three hours early, too." She turned her back to him. "Zip me up?"

He did so. "I'll pick you up for dinner later."

"Yes, you will."

Korolev opened the email's two attachments on the taxi ride back to his apartment. One was the call records for Trench's phone, and the other for Bozo's.

The first thing he noted was that there was no direct contact between the phones. At least Trench had had enough sense to use a throwaway.

A number on Bozo's list did catch Korolev's attention, though. It had been dialed just hours before the explosion. Korolev was sure he'd seen the number before.

He searched his contacts and found that it belonged to Pike Larson, an area arms dealer. He called the number.

"Hello?" The voice that answered was gruff.

"Mr. Larson, how are you today?"

"Who is this?"

"We've met before. Leonid Korolev. I work for the Greek."

"Mr. Korolev," Larson said, his tone suddenly solicitous. "What can I do for you?"

"Two days ago, you had a call from a man named Thomas Bozeman. You may know him as Bozo."

"Uh, yeah. He called. Why?"

"What did you talk about?"

"Hey, I respect you and the Greek and all you guys, but my work depends on confidentiality."

"I'm sure you're aware that Bozo is dead."

"I might have heard that."

"So, there's no confidentiality to be kept."

"Well, I, um . . ."

"Bozo did not die alone. The Greek's nephew was in the car with him."

"Wait! All I did was provide the device. I didn't set it off."

"Then it would be in your best interest to talk to me."

"What do you want to know?"

"Everything."

The rest of the story fell into place over the next few hours. When Korolev was confident he had all the facts, he called the Bean Counter.

Lauren answered the phone. "Hello, handsome."

"Good afternoon, gorgeous."

"Is this a social call or business?"

"Business, I'm afraid. I need to see the boss. Is he in?"

"Not at the moment, but he should be here in an hour."

"Can I see him then?"

"You can, or you can come now and wait until he gets here. I'm sure there's something I could find for you to do."

"I'm on my way."

"Trench killed himself with his own bomb?" the Bean Counter said. Gromyko was not going to like this.

"He did."

"How?"

"This Bozo character used some rookies to distract the driver of Trench's intended target, presumably to plant the bomb. Why it was still in the car Trench was in is a mystery."

"Did you find out who he was trying to kill?"

"At least two people, perhaps three. One is a woman he'd dated named Matilda Martin. The other is a lawyer she apparently took up with after Trench. There was a third person with them that night, not counting the driver, but I don't know her name yet."

"Trench's immature jealousy strikes again. Who was the attorney?"

"Stone Barrington. He's a partner at—"

"Barrington?" the Bean Counter said, cutting him off. "Are you sure?"

"Yes, sir. Why? Do you know him?"

"Of course, I know him. You should, too."

Korolev's brow creased.

"Barrington is the one who killed Serge Gromyko."

Korolev blinked. He'd been on an assignment in Las

Vegas when the first Greek had been killed, so he hadn't heard all the details.

"You're absolutely positive it's him?" the Bean Counter asked.

"One hundred percent."

"Dammit!"

Trench had been an idiot to the end. If he wasn't already dead, the Bean Counter would have killed him himself.

Things had finally calmed down where Barrington was concerned. With Alexei now in charge—and, in effect, the Bean Counter—business was running smoothly, and most importantly, profits were up. Any renewed thoughts of revenge against the attorney put all that at risk.

"What about the woman? Melinda . . . ?" he asked.

"Matilda Martin."

"What about her? Where is she?"

"From what I've learned, she's been staying with Barrington."

The Bean Counter cursed under his breath. "She must have her own place."

"She does."

"Put a team together and stake it out. She'll have to show up sometime."

"And when she does?"

"Bring her to me. And whatever you do, stay away from the lawyer. Do not cross paths with him."

If the Bean Counter played it right, he could use the woman as a sacrificial lamb to Gromyko and avoid the topic of Stone Barrington altogether.

"Yes, sir."

"And, Leonid, not a word about anything you've told me to anyone. Just tell them she is someone of interest."

"I understand," Korolev said.

He exited the Bean Counter's office and shut the door.

"Everything okay?" Lauren asked. "It sounded a little tense."

"Some news he wasn't expecting."

She wanted to ask him more, but she could get that from him later. She walked to him and played a finger down his chest. "Still on for dinner? Or should I look for other plans."

"Still on. I just need to set something up for the boss first, so I might be a little late."

She pushed up on her tippy-toes and kissed him. "You know it's not good to keep a girl waiting too long."

"I'll keep that in mind."

"See that you do."

Later that night, after Korolev had fallen asleep in her bed, Lauren tiptoed into the bathroom with her phone.

She had been involved in the organization for years, joining back when the Pentkovskys had been in charge, and had served as secretary to the youngest Pentkovsky, Egon, until the Bean Counter took over the role of mob CFO. And while she considered the Bean Counter a decent enough boss, she was less pleased with the Greek, both former and current.

In her mind, Egon—who now went by the name Peter Greco for safety reasons—should have been offered the top job instead of Gromyko. She had not given up hope of that happening one day.

To that end, she had been keeping him up to date on organization politics. It had been a week since she'd last contacted him, and a lot had happened in that time. Things that could directly affect him. Specifically, that the Greek was on the warpath in the wake of Trench's death. She thought it likely he'd take out his anger on the wayward Peter Greco. The Greek already considered him a threat, simply for being the brother of the family's first leader, Anton Pentkovsky. Now, with Peter slipping away from the family, who knew what the Greek might do?

She wrote a long text, and included the information that the Bean Counter had learned Stone Barrington was somehow involved with Trench's demise, but was holding off passing that information along to the Greek for the time being.

Once satisfied that she'd covered everything, she sent the text, then returned to the warmth of her bed.

CHAPTER 31

A week had gone by without an attempt on Stone's life. When there was not even the hint of anyone following him or paying him any undue attention, he called Mike Freeman. "It looks like I may have dodged a bullet. Figuratively speaking."

"I take it you remain untouched."

"Indeed. I think you can call off your men."

"Are you sure?"

"Gromykos have a notoriously hot temper," Stone said. "If the new Greek suspected I had anything to do with his nephew's death, I wouldn't be talking to you right now."

"I'm inclined to agree with you."

"Go put your men on a more exciting task."

"Very well. But if you sense even a hint of a problem, I can have them back in a hurry."

"Thanks, Mike."

The next day, Stone attended a meeting with Carly's new clients, Jones & Jones, in Bill Eggers's office, then walked downstairs with her, so she could show him her new office.

Stone looked around, impressed. "I had to move out of the building and into my house before I got an office this big," he said.

"That makes me feel good," she said, switching on MSNBC.

Stone walked two doors down to Herb Fisher's office and found him looking glum. "What's the matter, Herb?"

"Nobody comes to see me anymore. They all come to look at Carly and see if she's real."

"They'll get over it, then you'll be the star of the floor again."

"You promise?" Herb asked expectantly.

"Sure. I just came from a meeting with her new clients, in Eggers's office. Turns out they own two other businesses that are coming aboard, too."

"Please," Herb said, holding up his hands in surrender. "No more good news."

"Take the rest of the day off and go to the zoo," Stone suggested. Then he left the building and walked home.

When he arrived, Joan looked up from her desk, and said, "You look a little winded."

"It's called exercise," Stone replied.

"It's called heavy breathing," she said. She scrutinized him. "You aren't carrying, are you?"

"No," he admitted.

"I'm gonna tell Dino," she said, reprovingly.

"Don't you dare. I'm getting enough stick from him about that."

"You never believe that somebody's after you, until there's a bullet on the way, you know that?"

"I think a bulge in the armpit attracts more attention."

"It repels more attention," she said.

"You may have a point."

"Dino has the same point, and he's usually right."

"I hear enough of that from him. I don't need to hear more from you."

"Message received. There's a Mr. Greco waiting in your office. He didn't have an appointment, but his suit was nice, so I put him in there. Who is he?"

"Greco, you say? Not sure if he still does, but at one time he worked for at least one of the Greeks," Stone said.

"The Russians? Oh, shit." Joan took her .45 from her desk drawer. "I'll be ready," she said.

Greco was sitting in the chair across the desk from Stone's, drinking a cup of coffee. He was, indeed, wearing a nice suit—Savile Row, Stone surmised.

"Good morning, Egon," Stone said. Greco's original name had been Egon Pentkovsky, and his brother Anton had been the head of the Russian mob. Anton's demise had opened the door for Serge Gromyko, and then Alexei Gromyko, to take over. "Did Joan offer you something stronger?"

"Not before lunch," Greco replied. "And it's not Egon anymore. I go by Peter now. I thought if I was going to change one name, I might as well change both."

"Peter, then. I heard that Islesboro suddenly lost its attraction to you and your family." With Stone's assistance, Greco and his family had moved into a home on the island. Word had recently come to Stone that the Grecos had left Islesboro as quickly and quietly as they'd arrived.

"That is so. I wish I could have given notice, but there was no time to lose, given the position I was in at that moment. I thought I owed it to you to explain what had happened, but I had to wait to see you in New York."

"All right, shoot."

"Not the word I would have chosen," Greco said. "As you might remember, when I tried to disassociate myself from the family business, Serge Gromyko interpreted my actions as hostile to him, though I felt no hostility toward him, at that time."

"I recall. But didn't his passing solve that problem for you?"

"It did, in a way. But after his death, the new Greek contacted me and simply assumed I was still part of the operation. That's when I realized his brother had taken the knowledge of my betrayal to his grave. I was able to use the excuse of my own brothers' deaths to lessen my workload and move into more of a consultant role. I thought once I achieved that, I would be able to ease my way out completely, without causing any waves."

"Let me guess. It's not turning out to be so simple."

"That is an understatement. A week ago, I called him and told him that the time had come for me to branch out on my own. He did not take it well. As you might imagine, I have witnessed too much, and Alexei Gromyko does not tolerate witnesses. But that isn't the true problem."

"What is?"

"I am a Pentkovsky. My brother started the family. I now realize in Gromyko's mind I will always be a threat. If I were to stay in the family, he might not kill me right away, but he would eventually find a reason to do so. And I fear he'll do the same to my wife and daughters. I will not put my family at risk."

"So, you're still going to leave the mob?"

"I'm already out. They just don't know it yet."

"I see. But surely, he will be even more motivated to kill you than if you stayed. How do you propose to avoid the Greek's wrath?"

"Is our conversation being recorded?"

"Not unless I press a button, and I have not done so."

"You are the person who, along with me, has the most to gain from his exit from the planet."

"I am prepared to meet hostility with hostility, if it comes to that."

"It may very well come to that."

"You sound like you know something I don't."

Greco smiled without humor. "As you know, my original role in the organization was that of chief financial officer. I no longer hold the title, but I still have sources there who tell me things. And apparently a select few are aware of your involvement in the death of Gromyko's nephew."

Stone stiffened, but only slightly.

"So, it's true, then," Greco said.

"I will neither confirm nor deny."

"Lucky for you, that information has not been passed on to Gromyko. Yet. Not to put too fine a point on it, but when he does find out—and don't think he won't—your life will be at risk every time you walk out your door."

"I'm not sure you could put a finer point on it. Any notion on when the odds of my impending doom will increase?"

"If I were you, I'd assume they already have."

"So, you're saying I should never go outside again."

"I'm saying Gromyko is a problem that needs to be dealt with, because if he can figure out how to kill you in your chair, he will. Walls won't save you."

"Did you plan this visit to cheer me up?" Stone asked. "Which, incidentally, is not working. Or did you just happen to be in the neighborhood?"

"I was in the neighborhood, and I thought we should have a chat about our mutual problem."

Stone glanced at his watch. "Would you like to join me for lunch?"

"Thank you, yes."

Stone picked up the phone and pressed a button. "We will be two for lunch, in the study. Peter, do you have any dietary requirements?"

"None whatsoever. I eat anything."

Stone passed along the instructions and hung up. "Where were you born?" he asked, because he couldn't think of anything else to say.

"In St. Petersburg, then it was Leningrad, of course. We emigrated to the States when I was five. I became a citizen at sixteen."

"Your brother Anton was older?"

"By six years. He had already established the crime family by then. He made it clear that I might join him whenever I wished. I declined, but he didn't think my choice was permanent. He kept pestering me about it, until I acquiesced."

"And became his financial guy."

"Yes. A job that I continued to do for the Gromykos, for a while."

"Did you keep the books?"

"I did."

"Was there more than one copy?"

"I have a thumb drive containing everything for the past seven years. Including the time since the Gromykos took charge."

"That would be very useful to your not being killed by breaking free of the family."

"It's like you are reading my mind. The information I have would put Gromyko and many other influential members of the family away for good. But I can only do so if you help me get it in the right hands."

CHAPTER 32

They had omelets quietly, while Stone tried to figure out what to do with this new information about Peter Greco's thumb drive.

"Have you figured out what to do with the thumb drive?" Greco asked.

"Whoever you give it to, you should be represented in such a transaction," Stone replied.

"Will you represent me as my attorney in this matter?"

Stone thought about it for a millisecond. "Yes," he said. "Normally I would not, since we both have a stake in the outcome."

"Do we not have the same stake? Our lives?"

"Yes, so there is no conflict of interests."

"As my lawyer, what do you think we should do?"

"I expect that the Greek has committed crimes in Maine, New York, and, perhaps, in other cities and states.

We could pick one, but I think it's better, in this case, to go federal, if the evidence you have supports a federal crime."

"How about income tax evasion?"

"Very good. That's how Eliot Ness got Al Capone, when he couldn't get him for anything else."

"So, we go to the FBI?"

"Yes, but the right person at the FBI."

"And who might that be?"

"I might not trust most of the agents I have dealt with, but there is one who should be a straight enough arrow for this. His name is Thomas Kinder; he's assistant director for financial crimes."

"Whoever you wish."

Stone dug a throwaway cell phone from his desk drawer and dialed a number. "Assistant Director Kinder, please," he said to the person who answered.

A moment later, another man came on the line. "This is Assistant Director Kinder."

Stone put the phone on speaker. "Tom, it's Stone Barrington."

"Good day, Stone. What a surprise to hear from you."

"I have something that should go to someone at your level and not to the New York office."

"Well, I am at my level. Shoot."

"I have a client who is in possession of sufficient evidence of tax evasion and fraud to put Alexei Gromyko and many members of his organization in prison for a long time."

"So, you're playing Eliot Ness?"

"No, you are, if you want this case."

"In what form is the evidence?"

"It's on a thumb drive."

"And over what period of time were these crimes committed?"

"Seven years."

"And this witness is inside Gromyko's organization?"

"He was, until recently."

"And what does he expect for his cooperation?"

"The usual: immunity for any crime he may ever have committed and, very possibly, witness protection." Stone looked at Greco to confirm, which Greco did with a slight nod.

"And where is this witness now?"

"In New York City."

"Where in New York City?"

"In a safe place. You will have access to him as soon as you agree, in writing, to a fulsome statement of immunity, and an agreement that you, and only you, will deal with my client."

"This is an organization. He cannot deal only with me."

"You or those under your direct supervision," Stone said.

"Agreed."

Stone gave him his fax number and email address. "When can we expect you in New York?"

"Are you free for dinner?"

"Yes."

"Is the witness?"

"As long as dinner is at my house."

"Is seven o'clock good?"

"Just fine."

"I'll bring an associate for dinner, and two other agents will come along to handle video and audio equipment."

"I can feed them, too, but less sumptuously."

"They can rough it."

Stone gave Kinder directions on how to enter his garage from the avenue. "Your men can unload their gear in the garage. Don't arrive in vehicles with blue lights on top."

"Got it. We'll bring body armor, as well."

"Good. You may need it. I can put up you and your agent in my house. The tech people will have to find a hotel cheap enough to satisfy you."

"See you at seven."

"Make it six-thirty, if you want a drink before dinner."

"Considering who I'm dealing with, I'll stick to lemonade."

Stone called Joan next. "Please call Mike Freeman and tell him my request for his people to stand down was premature."

"On it," Joan said.

Stone hung up and turned to Greco. "We're all set. Will it upset your wife if you sleep here tonight, instead of going to wherever home is at the moment?"

"That will be fine."

"I can have someone pick up some clothes for you."

"Suite One, Waldorf Towers. I'll phone my wife and ask her to pack a bag for me."

"Okay. I'm going to ask a young woman, a lawyer at

my firm, to join us. She has a prodigious memory, so we won't have to take notes. I'll get us a copy of whatever the FBI records, too."

"Sounds like you've thought of everything," Greco said.

"I hope to God that's true," Stone replied.

Stone phoned Carly Riggs and asked her to come to his study at six o'clock. "This will be a business dinner," he said to her.

"And what do you require of me at this dinner?" she asked, when she was seated in his study.

"I require you to remember whatever is said, and to prepare a transcript from memory."

"Who are the attendees?"

"Thomas Kinder, assistant director of the FBI for financial crimes, and an agent of his choosing. There will also be two technicians hovering about to make audio and video records of what transpires."

"Who's the client?"

"A new one: Peter Greco, né Egon Pentkovsky."

"Of the Russian mob Pentkovskys? Brothers, right? Anton Pentkovsky was in charge prior to Serge Gromyko."

"Well, Peter's the brother who's still alive. He served for seven years as the organization's chief financial officer. He believes that Alexei Gromyko, the mob's current leader, wishes him dead, so he is offering his financial records, contained on a thumb drive, to the FBI, in return for immunity and witness protection."

"Okay, when do we start?"

"As soon as the FBI men arrive, we'll have dinner. Then I'll produce Greco, and he will disclose all."

"How long will the meeting take?"

"As long as it takes. We may have to pause and continue tomorrow. Everybody's sleeping here."

"Okay, I'm game."

"Good."

Kinder and his associate, Bob Grant, arrived, declined alcohol, and had dinner. Then they got down to business.

"So, where is your mysterious witness?" Kinder asked.

"Upstairs. He recently changed his name to Peter Greco, but you will probably know him by his birth name, Egon Pentkovsky."

"Anton and Izak's brother?"

"Do you know of any other?"

Kinder raised an impressed eyebrow, then turned to Carly. "Ms. Riggs, what is your purpose at this meeting?"

"To remember it," Carly replied.

Kinder looked at Stone, askance.

"Carly has an extraordinary memory," Stone said. "When this is over, you will wish to hire her, but you may not. She belongs to Woodman & Weld."

"Okay," Kinder said, and there was disbelief in his voice. "We are all set up in your study. Produce your witness."

Stone picked up a phone and dialed an extension. "Now, please."

They moved to the study, and the subject turned up. Introductions were made, and everyone took a seat.

Stone handed Kinder a document outlining what Greco was receiving for his testimony. "Read this and sign it."

Kinder read it and, to Stone's surprise, signed it without requesting alterations.

"To what extent will Mr. Greco require witness protection?" Kinder asked.

Greco spoke up. "I will choose a location, and you will transport me there and provide security for as long as I deem necessary. I will not require that you furnish housing or create a new life for me. I'll take care of that myself."

"Stone," Kinder said, "I believe your client is underestimating the extent of our involvement in his safety."

"I assure you, I am not," Greco said. "I have already purchased a property in a western state under a corporate name. It will be easily surveilled and guarded by your people, and housing will be provided for them."

"As you wish," Kinder said. "Let's get started."

Five hours later, they concluded the meeting. Everyone at the table appeared very tired.

"That was the most productive such meeting I have ever attended," Kinder said. "I will need tomorrow to review it and compose other questions, if necessary. After I've done that, it will be time to involve a United States attorney to assess the chances of success at prosecution. May we meet again tomorrow evening?"

"Of course," Stone said.

Kinder and Grant had a brandy, then repaired to their rooms for the night.

"You did astoundingly well," Stone said to Greco.

"Thank you."

"Carly, did you get everything?"

"Of course," she said. "I'll dictate it to Joan tomorrow, and she can type it up."

They adjourned for the evening.

CHAPTER 33

They met again in Stone's study the following evening, and this time Tom Kinder and his associate had a drink.

"Where's your U.S. attorney?" Stone asked.

"She will be here shortly," Kinder replied. "I'm told that when asked to provide a prosecutor, she volunteered herself."

A distant alarm bell went off in Stone's brain. But before he had time to identify it, a tall, beautiful blonde walked into the room.

"Stone, may I introduce Tiffany Baldwin, the United States attorney for the Southern District of New York? Tiffany, this is Stone Barrington, attorney for our witness, whom you have already met, and his associate, Carly Riggs."

Stone's heart sank. He and Tiffany had had half a dozen

rolls in the hay in years past, and she would not let go. He avoided her at all times.

"Oh, yes," Tiffany replied. "Stone and I are old friends. How do you do, Carly?"

"I do very well, thank you," Carly replied. She appeared to have immediately sensed the circumstances of Stone's previous acquaintance with Tiffany.

"Shall we start?" Stone asked. He conducted the meeting without so much as glancing at Tiffany, to Carly's evident amusement.

When they were done, they all had a nightcap. Then Greco departed with the FBI, while the U.S. attorney didn't budge from her seat. Neither did Carly.

STAY! Stone mouthed at Carly, who kept a straight face.

"May I have another drink?" Tiffany asked.

Stone poured her a very small Scotch.

"Carly," Tiffany said. "Stone and I have something to discuss. Would you excuse us, please?"

"I'm sorry, Tiffany," Stone said, "but Carly goes where I go this evening. I can't have meetings with opposing counsel without her present."

"You want a witness, then?"

"I think that will help keep the conversation on track."

"I'll be as quiet as a mouse," Carly said sweetly.

"This is a rather intimate discussion," Tiffany said.

"I have nothing to hide from Carly," Stone said quickly.

"Looks like I'm here for the duration," Carly said, handing Stone her glass.

Stone poured her a hefty drink and handed her back the glass.

"Shall I close my eyes?" Carly asked, innocently.

"No," Stone said firmly.

Tiffany appeared to be ready to spring at him.

"Do tell me how you and Stone happened to meet," Carly said.

"It was some years ago," Tiffany replied through clenched teeth. "We were on opposite sides of the bed, in those days."

"And still are," Stone said.

"This is so much fun," Carly said, laughing.

"I'm prepared to make it less fun, if you don't leave," Tiffany said to her.

"Please, Tiffany," Stone interjected. "Carly is both my business associate and my guest."

Carly smiled. "While you are simply Stone's opponent in a case," she said. "Tell me, how long has it been since you and Stone have, ah, shared each other's company?"

"Too long," Tiffany replied.

"Not long enough," Stone said simultaneously.

Tiffany glared at Stone. "Well, perhaps I should leave and allow you and this, ah, girl, to get on with it."

"Thank you so much," Carly said.

Stone pressed a button and, momentarily, Fred appeared at the study door. "Fred," Stone said, "would you kindly show Ms. Baldwin out?"

"I'll see you tomorrow evening, then?" Tiffany said, smoothing her skirt.

"No," Stone said. "Tom is returning to Washington tomorrow morning, so our business is concluded."

"Surely, there are things we haven't covered," Tiffany said.

"No, we are all done," Stone said. "Good night, Tiffany."

"Yes, Tiffany," Carly said. "Good night." She waited until she heard the outside door close before she spoke again. "Oh, good. I was afraid I'd have to help you fight her off, Stone."

"Your fears were well-founded."

"An old flame? Or perhaps, torch?"

"And one difficult to stanch," Stone replied. "I made the error once of giving her a key to the house. I had to have the locks changed."

"Well, that was a mistake."

"You have no idea."

"Then I bid you good night," Carly said, rising. "Call me, if you need reinforcements." She got up and left.

CHAPTER 34

For several days, Korolev had a team staking out Matilda Martin's apartment, but she hadn't shown up. After consultation with the Bean Counter, he had moved his team to Stone Barrington's neighborhood, in hopes of spotting her. This proved equally unsuccessful.

He had learned the house next to Barrington's was also owned by the lawyer, and that he often let friends live in its apartments.

"She's either in one of the apartments or in Barrington's main house," he said to the Bean Counter.

"Barrington's house is off-limits."

"I understand that, but what about the apartments?"

"Didn't you say there were security guards?"

"Only the first night. None since then."

They'd discussed it for a bit before Korolev finally

convinced the Bean Counter to let him check the apartments. Tonight was the night.

The lights in the lawyer's house and the house next door finally went out.

Korolev lifted his radio, and said in Russian, "Standby."

Waiting down the street were two of his men—Ruddy Antonovich and Theo Malic, Korolev's best lock picker. They would be the ones searching the apartments while he kept watch from across the block, waiting in the stolen sedan they would use as a getaway car.

He let another five minutes pass, and when the lights in either house remained off, said, "Now."

His men moved up to the entrance of the apartment building. In a handful of seconds, the door swung inward, and the pair hurried inside to the door of the nearest apartment.

Ruddy pressed his ear against it. All was quiet. He nodded to Malic, and the man once more did his magic.

The apartment was dark. From deeper inside came the soft sound of steady breathing. Ruddy followed it to a bedroom where he found a woman asleep. She didn't look like either of the women Korolev had described, so he quietly returned to the hallway.

The next apartment was unoccupied, and they quickly moved on to the third.

When Malic eased the door open, they could see a dim light coming from farther inside.

Ruddy pulled out the special spray can Korolev had given him and crept through the flat, following the sound of running water to a bathroom.

Inside, a woman was brushing her teeth at the sink. She matched one of the descriptions; the problem was, he couldn't remember if it was Matilda Martin or the other one. She was a real looker, though. Just the kind of woman Trench had gone for. So, she had to be Matilda.

Korolev's voice came over the radio again, shouting in Russian, "Get out of there! Lights are back on in the main house."

The woman spun toward the door. "Who the hell are you?"

Reflexively, Ruddy raised the can and sprayed her in the face.

She coughed and sputtered, "What the . . ." then crumpled to the floor.

On the radio, Korolev said, "Did you hear me?"

Ruddy pressed the talk button. "Yes. I found her, boss."

"Where is she?"

"Lying on her bathroom floor. I sprayed her."

"Okay, okay. Good. Get her out of there. Quick. I'll bring the car around."

Ruddy carried her over his shoulder to the building entrance, with Malic right behind him. The moment Korolev drove up, they hurried to the sedan, and got in with the woman.

Once they gone far enough that Korolev felt sure no one had followed them, he parked on a quiet side street, and turned to the back seat.

"Let me see her face," he said.

Ruddy moved away her hair, and Korolev stared in disbelief. "That's not her!"

"What?" Ruddy said.

"Did I stutter?"

Ruddy looked at the unconscious woman. "She fits the description."

"Yeah. Of the *other* woman. That's not Matilda Martin. Fuck, Ruddy. What's wrong with you?"

"I'm sorry. I—I got mixed up, I guess."

"You think?" Korolev ran his fingers through his hair.

"What should we do with her, then?" Ruddy asked.

"Good question." Korolev glared at him. "What do you think we should do?"

"I don't know. I guess we could kill her and dump her somewhere."

"That all depends. Did she see you before you sprayed her?"

Ruddy realized if he said yes, he'd be in more trouble than he already was. "Uh, no. She . . . she never saw me."

Korolev studied him. "You'd better be telling me the truth."

"I am," Ruddy said.

"So," Malic said, "what are we going to do with her?"

"Not make this an even bigger mess than it already is," Korolev said, then pulled the car from the curb.

CHAPTER 35

The *Times* and the *Post* arrived with breakfast the following morning. Both hinted that a big-time arrest by the feds was in the offing. Tomorrow, it would be headlines, Stone thought.

Joan buzzed. "Alexei Gromyko on one for you."

Stone picked up, trying to sound casual. "Hello?"

"Have you done something that can't be undone?" the Greek asked.

"I don't know what you mean," Stone said, keeping his voice steady. He wasn't sure if Gromyko was talking about Trench or Greco or both.

"I believe you do."

"Mr. Gromyko," Stone said wearily, "do you have a point to make?"

"When I do, you'll know it," Gromyko said, then hung up.

Stone immediately called the Waldorf Towers and asked for Peter Greco.

"I'm sorry," the operator said. "That party checked out early this morning."

"Thank you." Stone hung up and called Tom Kinder.

"Assistant Director Kinder."

"It's Stone. Have you heard from Peter Greco this morning?"

"No, we dropped him off at his suite door last evening. I had intended on leaving agents to guard his room, but Greco worried that would bring more attention to his presence. We compromised with leaving a couple agents in the lobby."

"Then I take it your agents didn't see him check out early this morning."

"No, they did not," Kinder said. "I'll send my agents up to have a look at his suite. I'll get back to you." He hung up.

Joan buzzed again. "Herb Fisher on one for you."

Stone picked up. "Good morning, Herb."

"Not yet. What have you done with my associate?"

"Nothing at all."

"She hasn't shown up for work this morning, and she didn't call. Normally, she's meticulous in her work habits."

"I saw her last evening when we finished work. That was about nine PM. I'll see what I can find out."

He buzzed Joan.

"Yes?"

"Carly is MIA. See if you can find her."

"Righty-o." Joan hung up.

Stone buzzed Fred.

"Yes, sir?"

"Have you seen Carly this morning?"

"No, I didn't drive her anywhere."

"Is her car still in the garage?"

"It is," Fred said.

Stone played at working for another hour, then the phone rang.

"Assistant Director Kinder on one."

Stone picked up. "Tom?"

"My guys went through Greco's suite with a fine-tooth comb. Nothing to indicate foul play, or anything else."

"Carly Riggs is out of pocket, too. She should be at Woodman & Weld, but she's not. If either she or Greco were missing, I'd be concerned. Right now, I'm starting to feel panicky."

"I'll put out an APB for our New York office, but that's not a lot of people on the street."

"I'll get back to you," Stone said. He hung up and dialed Dino.

"Bacchetti."

"I've got two missing persons on my hands," Stone said.

"Anybody I know?"

"Peter Greco and Carly Riggs. Greco is turning state's evidence against Gromyko, so I doubt the Greek is happy the FBI is looking for him. That bit is confidential."

"I heard nothing. You want an APB on both?"

"Yes, please."

"I'll call you back." He hung up.

Joan was standing in the door. "I haven't been able to locate Carly anywhere."

"Go through her apartment and see if you can find any sort of communication with us—notes, lipstick on the mirrors, anything."

Joan walked in fifteen minutes later. "Nothing," she said. "But Carly has four very nice handbags; I saw them when she moved in. They're all in the apartment, and she wouldn't have left the building without one of them."

"Why not?"

"Because she's a woman, and women don't go anywhere without a handbag. Oh, and one of them had her wallet in it, and some money. That's likely the one she intended to take with her."

"Dino is on it," Stone said. "He's the best we can do."

"I'll wait for his call," Joan said.

An hour later, Joan buzzed. "Carly on one."

Stone yanked the receiver off the hook. "Carly, where are you?"

"On the corner of Fifty-Seventh and Third Avenue," she said. "I had to borrow someone's phone."

"How did you get there?"

"A man came into my apartment last night and sprayed

something in my face. I woke up about five minutes ago behind a dumpster and went to the street."

"Do you see any cabs from where you are?"

"Yes! Be there soon." She hung up.

Stone buzzed Joan. "Carly is on the way here in a cab. Keep watch for her, and go armed, just in case."

"Gotcha!"

Stone grabbed a weapon from his desk drawer and ran to the front door of the house. He watched through the peephole, in case she got out there. Nothing.

Stone called Joan's cell.

"No sign of her."

"Here I am," Carly said.

Stone turned to find her standing in the door of his office.

"I'm so relieved to see you," Stone said.

"I'm relieved, too," she replied.

He buzzed Joan. "Go pay for that cab out front and call everybody off the search for Carly," he said.

Joan appeared in the doorway, and Stone held up a hand to stop her. "What happened?"

"I was kidnapped," Carly said. "They hit me with some sort of spray and put me out. I woke up in an alley near Bloomingdale's."

"I want some of that spray for my handbag," Joan said.

"Any idea who they were?" Stone asked.

"I heard a voice over a radio speaking in Russian, right before I was knocked out."

"Anyway, she's safe," Stone said. "Carly, do you feel like going to work at Woodman & Weld or would you like to rest?"

"Work sounds good right now."

"Joan, call Herb Fisher and tell him Carly is on the way in, and not to give her a hard time about being late."

"Right." Joan disappeared.

"All right, listen carefully, Carly. You are not to leave this house or Woodman & Weld, except in my company or that of Fred in the Bentley. Got that?"

"Got it."

Joan came into the office, waving an envelope. "Carly's carry license came, delivered by a uniform." She handed it to Carly.

"Good timing. Find her a .380, a couple of spare magazines, and a box of cartridges." Joan nodded.

Carly opened the envelope and stared at the license.

"That is a license to carry, not a license to kill," Stone said. "If you kill somebody who is not shooting at you or strangling you at the time, your life will change forever, and for the worse."

"I understand all that," she said. "You've told me before. Could I have shot those two guys who took me last night?"

"Only while they were engaged in the taking of you."

"Okay, point taken."

"That being said, carry it everywhere. Joan, find her that catalog of holsters, and let her pick out something. Some ladies prefer the Thunderwear."

"The what?"

"Joan will explain that to you. Joan, call Fred, brief him, and ask him to take Carly to work, but not until her weapon is loaded and available."

Joan beckoned Carly to follow her, and they left.

The phone rang, and Stone grabbed it. "Yes?"

"It's Herb. What the fuck is going on?"

"Carly will be on the way soon, and she'll explain it to you. She's going to be traveling armed now, so watch yourself."

"Is she going to shoot me?"

"Not if you're careful of her."

"Don't worry, I'm not going anywhere near her."

"That's a good policy." Stone's cell phone rang. "Gotta go." He picked up the other call. "Yes?"

"Stone, it's Peter Greco."

"Peter, Thomas Kinder has been looking everywhere for you. Are you all right?"

"I'm fine, and I just talked to Kinder. I saw a face I recognized near the Waldorf and thought it best if my family and I found alternative lodging right away."

"Smart move. Where are you now?"

"Long Island, at the home of someone I trust. But I'd rather not stay here long. The family has business interests in the area. You helped us once before, and I was hoping you could do the same again."

"Would you consider a return to Islesboro? Or do you think Gromyko would suspect that?"

"I think that would be perfect. I doubt the Greek will look there."

"Let me make a few calls and get back to you."

They got off the phone, and Stone contacted Jimmy Hotchkiss, who knew everything about property on Islesboro. Jimmy promised to forward information to Stone by the morning.

Stone's phone rang again.

"It's Dino. I hear there's a happy ending."

"No, a happy ending would be if Gromyko were dead."

"You think he was behind the disappearances?"

"Carly was kidnapped last evening. She said they were Russians, and he's the number one Russian around here. Thanks for your help with her carry license. She's carrying now."

"Glad to hear it. Any word on Greco?"

"He's safe and sound, so you can call off the dogs."

"Did he say where?"

"Only in vague terms, which is probably for the best."

"Agreed."

"He is also out of your jurisdiction."

"The news keeps getting better and better. Then, if everything's settled, I should probably do some work."

"The taxpayers would appreciate it."

"See ya." Dino hung up.

Joan came back in with Carly. "Okay, she's ready to go. The Thunderwear is being overnighted."

"Don't either of you shoot the FedEx guy," Stone said. "Fred awaits. Carly, remember: no leaving the house or the office without him, me, or both of us."

She tapped her head. "And you remember, I never forget anything."

"I'll take her to the garage," Joan said.

After they left, Stone sat down and took some deep breaths.

CHAPTER 36

The speaker on Lauren's desk buzzed, followed by the Bean Counter saying, "Send him in."

She looked at Korolev, not hiding her concern.

He gave her a smile that said everything would be fine and then headed to the door. Everything was not fine. He had been dreading this meeting since Lauren had woken him with a call at nine AM, telling him he was expected at eleven.

He entered his boss's office.

The Bean Counter was standing behind his desk, pulling his suit jacket on. Korolev stopped a respectful distance away and waited.

The Bean Counter checked his computer screen and picked up his briefcase before finally looking at Korolev. "Let's go."

"Go? Where, sir?"

"To see the Greek."

Korolev felt a chill run down his back.

The Bean Counter didn't say another word until they were in the back seat of his sedan, on the way across town. "I assume you failed."

"It did not go as hoped."

"What happened?"

Korolev told him about last night's attempt to kidnap Matilda Martin, leaving nothing out, knowing to do so would not go unpunished.

"Are you even sure she was there?" the Bean Counter asked.

"I . . ."

The Bean Counter raised an eyebrow.

"No," Korolev said. "I'm not sure. She has not been seen for several days."

"Is it possible she left the city?"

Korolev reluctantly nodded. "It is possible."

The Bean Counter sighed. "Then I guess I have no choice."

The Greek's assistant met the Bean Counter and Korolev at the entrance of a building in Brooklyn that was undergoing renovation and led them through the first floor to a closed door.

"One moment, please," the assistant said, then went inside. Five minutes passed before he reappeared and motioned for them to enter.

The Greek sat at the only table in the room, working on a plate of *cacio e pepe*. In three of the other chairs were members of his inner circle. None of them were eating.

Without looking up from his food, Gromyko said, "Nice of you to stop by. I haven't seen you for what? More than a week now, isn't it?"

"My apologies, Alexei," the Bean Counter said. "I've been busy."

"You must have been. Can I assume the reason you're here now is because you finally have an answer for me?"

The Bean Counter had hoped to avoid this, but the failure to find Matilda Martin and offer her up instead meant he no longer had a choice. "I do. I would have reported to you sooner, but I wanted to make sure there was no mistake."

Gromyko looked at him. "So, tell me. Who killed my nephew?"

"The short answer is he did it to himself."

The Greek stared at him, face blank.

"The bomb was his," the Bean Counter said. "He intended it for someone else, but it went off when it was still in his possession."

His boss's gaze did not change.

The Bean Counter gestured to his companion. "I asked Mr. Korolev to learn all he could about the incident. He can fill in the details."

Gromyko's eyes shifted to the younger man without his face moving. "Leonid, isn't it?"

"Yes, sir," Korolev said.

"I've heard good things about you. Prove to me I have not been misinformed."

Korolev told the Greek everything except about the failed mission the night before, as it had no bearing on the question that had been asked.

At the mention of Stone Barrington, the already uncomfortable atmosphere turned tense.

When Korolev finished, the Greek focused his attention back to the Bean Counter. "How long have you suspected Barrington's involvement?"

"His involvement is limited to being an acquaintance of a woman Trench was interested in."

"How long?"

"A few days."

"And you didn't think it was important to tell me immediately?"

"I wanted to determine his role, if any, in what—"

"Stop." Gromyko closed his eyes and took a couple of deep breaths, then pushed himself to his feet. Teeth clenched, he said, "Barrington is responsible for the deaths of at least the last two heads of this family, and dozens of our men. If he even breathes in the direction of one of our people, I should be told. Or do you not think that's the case?"

"I didn't want to give you—"

"I don't care what you wanted. What *I* want is Barrington's head on a pike, mounted to the door in front of his law firm."

He walked up to the Bean Counter, until their faces were only inches apart. "Are you even aware that Barrington is already involved in an even bigger problem for us? Or was your head too far up your ass hiding this from me to notice?"

The Bean Counter's brow furrowed, unsure what the Greek was talking about.

"Why do you think we're meeting here in this dump?" the Greek said. "Because your buddy Stone Barrington has convinced the FBI to arrest me. There are agents scouring the town looking for me right now. Were you aware of that?"

"I, uh . . . no."

"Of course, you weren't. By not telling me until now, you have put me at a disadvantage. I do not like being at a disadvantage."

He turned back to the table and swept his plate onto the floor, then with his back to everyone said, "I want Barrington dead. Now."

CHAPTER 37

Joan buzzed Stone. "Lance on one."

Stone picked up the phone. "Why do you bother being announced?"

"Sometimes it's convenient."

"What can I do for you?"

"I want to meet Carly Riggs."

"Why?"

"She sounds interesting."

"She *is* interesting. That's why you can't meet her."

"Don't be that way, Stone."

"You're recruiting, and she's not currently available."

"When will she be available?"

"After she's made partner at Woodman & Weld."

"That's too long to wait."

"She's done Yale, Yale Law, and she aced the bar exam."

"That's why I want to meet her."

"She has a career in the law planned, and I don't want you to screw it up for her by shipping her off to the Farm, then to God-knows-where." The Farm was the CIA's training facility in Virginia.

"That would be a good career for her."

"It would be good for you, not her. Carly needs protecting from all that for a while, maybe for a long while."

"Why don't we let her be the judge of that?"

"Because I don't know yet how good her judgment is."

"We like them when they're still a little malleable."

"Lance, I can't stop you from trying, but I'll do everything I can to keep her where she is and doing what she's doing."

"Can you protect her from the Russians?"

"Yes." Stone wished he were as confident as he sounded.

"We can protect her at the Farm."

"We can protect her here, too."

"You didn't protect her last night," Lance said.

Why did Lance know everything? "A minor slipup. It won't happen again."

"And how will you prevent that?"

"By shooting everybody who tries."

"You might be able to trade her for peace with Gromyko," Lance said.

"I already have peace with Gromyko. He's being arrested this morning for income tax evasion."

"Are you sure about that?"

"I'm expecting a call momentarily."

"He'll just bail out and disappear."

"He won't get bail."

"You think he can't get bail out of Tiffany Baldwin?"

"Yes."

"And what happens if Tiffany doesn't play?"

"I'll shoot her myself," Stone said.

"Bravado doesn't suit you, Stone."

"It's courage, not bravado."

"What's the difference?"

"Content of character."

"You don't have it in you to shoot anybody."

"I've done it before. I'll do it again."

"Face-to-face? In the head?"

"Twice."

"Are you going to shoot me, if I recruit Carly?"

"Not you."

"Why not?"

"Because, if I shoot you, there are too many people who would care. Nobody cares about Gromyko."

"That's true, I suppose. All right, I'll give you a little more time with Carly."

"You'll stay away from her at all times, or you'll make me mad," Stone said.

"And why do you think I would fear you mad?"

"Because, if you think about it, you'll realize how many ways I could screw you."

"Such as?"

"Do you like your job, Lance?"

"Of course."

"Would you like to keep it?"

There was a silence. "You're threatening me with going to the president?"

"If I do, you'll have a hard time working anywhere, ever again."

"That's a serious threat," Lance said. "Can you back it up?"

"Try me."

Another long silence, then Lance hung up.

Stone hung up, too, and his hand was trembling. Nobody had ever talked that way to Lance. Maybe he had gone too far.

CHAPTER 38

When Carly came down to the lobby of the Seagram Building, Stone was waiting for her in the Bentley.

"This is unexpected," she said, sliding in the back seat next to Stone.

"It's necessary," Stone said.

"Is Gromyko in jail?" she asked.

"I'm still waiting for that call. In the meantime, there's another threat to deal with."

"Now what?"

"Lance Cabot."

"The head of the CIA?"

"That's the one."

"What's the problem with him?"

"He's taken an interest in you."

"What sort of interest are we talking about? Sex?"

"No, nobody's sure whether Lance does that. He wants to recruit you for the Agency."

"Well, that's flattering. Does he always do that personally? I thought he had a nationwide network of recruiters at such places as Ivy League universities."

"Oh, he does, but he always has an eye out for talent, and when he spots it, he wants it. And he's spotted you."

"Is he so hard to say no to?"

"He is, and there's an element of danger in that."

"How so?"

"Here's how it works," Stone said. "First, he'll take you to a very good dinner, and he'll talk the whole time. You won't even have a chance to ask questions, until he's finished."

"Okay, I can handle a good dinner. What then?"

"Then he'll take you down to a place called the Farm, in Virginia."

"I've read about that; it sounds interesting."

"Lance thinks you'll think that, especially when he tells you about the kind of training you'll get there."

"For how long?"

"The standard is about twelve weeks, but he would likely pull you out before that."

"Why?"

"Because he'll have a special assignment that your training has shaped you for, and it will sound enticing."

"Okay, then what?"

"From then on, Lance will keep you very busy. You'll visit exotic places and do work that you feel might just save the planet."

"Sounds like a dream job."

"It is, in its way, but it's dangerous. Survival is a very important part of the work."

"Why do you know all of this?"

"Because I have attended the Farm, albeit briefly, and I know others who have done so under Lance's watchful eye."

"Like who?"

"One of them is currently serving as president of the United States."

"Holy shit! Sounds like I'd enjoy some of that training."

"You will, some of it. But you're no Holly Barker; no one is."

"I'll try to hold my ambition in check."

"Later in your career, it might be worth your while to study under Lance, but not yet. You're going to have a lot of options in your life, and the Agency is only one of them."

"Well, that's flattering."

"We're not talking about flattery, here, Carly. We're talking reality. You're going to have a lot of fun doing the things you're still learning to do, but believe me, it's too early in your young life for you to learn it at the hands of Lance Cabot."

"Okay, say Lance approaches me. How do I handle it?"

"You give him your undivided attention. Never look bored or inattentive, then while he's talking, think about why you don't want to do what he's suggesting you do.

When he's done, tell him no, in specific, but polite terms. And be firm in your resolve."

"What if what he's offering is really, really enticing?" she asked.

"Trust me: you *don't want to do it.* You should believe me, because I have your interests at heart, while Lance has only *his* interests at heart. There's nothing I want for you but to do well and make good choices."

"I believe that, Stone," she said. "And I'll do as you ask."

"Good. After Lance has made his pitch, we'll talk about it, and you'll see why it's not a good idea."

"All right."

"Home, Fred," Stone said. "And without getting shot at, if you can manage it."

"I'll do my best, sir."

CHAPTER 39

Stone was dying to know what had happened in the Gromyko case, but he was afraid to call Tiffany Baldwin and ask her. Finally, he called Thomas Kinder, at the FBI.

"Hello, Stone," Kinder said.

"Hi, Tom. What's happening with Gromyko? I was afraid to call Tiffany Baldwin."

"You're scared shitless of her, aren't you?"

"You would be, too, if you'd had my experience with her. And don't ask! So, is the Greek in jail?"

"Not as of yet."

"What's the problem?"

"When my men went to arrest him, he wasn't there, and we haven't been able to find him."

Stone gulped. "Is he still in New York?"

"I don't know. I said we couldn't find him, remember?"

"He's keeping a low profile. That would explain why his people were able to kidnap Carly Riggs yesterday. He grabbed her right out of my house, sprayed something in her face that rendered her unconscious."

"*That's* what happened to her? I'll see if my people have found anything about where she is yet."

"Not necessary. She came to behind a dumpster and managed to get away. She's safe now."

"Thank God! You gave me a turn there."

"Imagine the turn I got," Stone said.

"Well, yes."

"Have you alerted Dino about Gromyko?" Stone asked. "He can manage a very big net with a phone call."

"Ah, no. I guess I didn't want to explain why we hadn't had an arraignment."

"I'll alert him."

"I'd be grateful, and I will keep you posted, Stone." He hung up.

Stone called Dino.

"Bacchetti. Don't tell me, Gromyko hasn't been arrested."

"How did you know?"

"Because I run the kind of shop where, when people say to me, 'I'll get back to you,' they get back to me."

"I just talked to Tom Kinder and heard only discouraging words."

"Does he want me to put an APB out?"

"He does, but he doesn't want you asking questions."

"I'll take care of it."

"As I knew you would."

"Shall we take Carly to dinner tonight?"

"I'll ask. Clarke's at seven-thirty?"

"Done." Dino hung up.

Dressed for dinner, Stone came down from his room a little early to find Helene waiting at the bottom of the stairs. Carly was with her, wearing something new from Bloomie's.

"Mr. Barrington," Helene said, "Fred asked that you leave by way of the garden gate."

"And why is that?" he asked.

"He says the Strategic Services people spotted two suspicious people in a car parked down the street, watching the house."

"That's a good reason."

"He said he will pick you up there, in an SUV he borrowed from the firm."

"Thank you, Helene."

She smiled and left.

"Shall we?" Stone said to Carly.

As they made their way to the gate, she asked, "Do you think it's the Russians?"

"As far as I know, no one else has the immediate desire for my death."

"Maybe we should go in separate cars."

"Carly, I'm shocked. By now I would have thought

you knew the best place to be is closer to me, not farther. Though many have tried to get rid of me, none have succeeded."

"I would like to see the statistics on those around you when prior attempts have been made, before I pass judgment."

Stone patted his jacket pockets. "Sadly, I seem to have left that information in my office. If it helps, I trust Fred has everything well in hand."

"That will have to do."

Fred was right where he'd said he would be, and soon they were on their way.

Stone called Mike Freeman. "I'm not sure if you heard yet, but there are a couple of unwanted admirers camped out in a car, near my house."

"I just received the report," Mike said. "Gromyko's people?"

"That would be my assumption. I think it's time to do what we talked about." That afternoon Stone had called Mike and discussed the possibility of beefing up the security detail watching the house, and providing bodyguards for when Stone went out.

"I thought you'd say that. I've already contacted additional resources and they will be in place within thirty minutes. If that doesn't get Gromyko's people to leave, we'll take more direct action. Are you home now?"

"No. I'm on the way to Clarke's."

"I'll have another car meet you there."

"Thank you, Mike." Stone hung up.

Stone and Carly were first to arrive and were shown directly to their table and placed their drink orders. Stone had had an idea that day, and asked, "Carly, do me a carlysearch on your memory."

"For what subject?"

"Teddy Fay." He spelled it for her.

"Got him," she said. "Former CIA tech expert, with an unproven reputation as an assassin."

"I'm thinking of asking him to join our little anti-Gromyko campaign."

"Well, if the rumor that he is still alive is more than a rumor, sure, why not?"

"He has an alias," Stone said. "Billy Barnett."

"Producer at Centurion?"

"You are quick."

"He was mentioned in a review of your son, Peter's, recent film."

"And that was enough for you to store his memory?"

"I don't have a lot of control over what sticks in my memory. Teddy Fay did. And Billy Barnett, too."

"Need I tell you that, should you meet, you are never, *never*, to speak his real name aloud?"

"Okay, got it."

"For our purposes, your memory is now a vault, and only you have the combination."

"Agreed."

Stone checked his watch. "He should be here momentarily."

A man appeared at their table.

"Your martini, *m'amselle*," the man said, removing the drink from a silver tray.

"Carly Riggs, may I introduce Billy Barnett?"

"How do you do?" Carly said. "And how did you . . ."

"A waiter was already approaching with your drink," Billy said and sat. "I'm not a real magician."

"Then who choreographed this little scene?" she asked.

"Stone did."

"All the elements for a flush were at hand," Stone said. "I just dealt the cards."

"From the bottom of the deck," Carly said.

"I believe you intentionally asked Dino to come half an hour later," Billy said.

"That's right. Shall I begin?"

"Please." Someone set a Knob Creek before him, and before Stone, as well. "Shoot."

Stone delivered a three-minute disquisition on Alexei Gromyko. "Have I left out anything?"

"You left out me," Carly said.

"I filled in Billy on the subject of you this afternoon."

"And you got here from L.A. in time for dinner?"

"I was in Miami, on the way to the airport for a flight home. I canceled that one, and I caught the next plane to JFK."

"Of course."

"People always underestimate Billy, Carly. It's a trait you might adopt for yourself."

"I already have," she replied.

"Then you might leave out a few chapters, next time.

You don't need to be completely knowable at first glance," Billy said. "There's value in mystery."

"Tried it, and it didn't work for me." she replied. "There's a lump of compulsion at the center of my psyche."

"And you must learn to control it, now and then."

"Okay," Carly said to Stone. "He's smart; smarter than I."

"You used the proper pronoun in that sentence," Billy said. "It's a start."

"Finally, I scored," Carly said.

Dino arrived.

"Good evening, Carly. And, Billy, nice surprise."

"Thank you, Dino," Billy said. "I've been getting to know Carly."

"Good luck with that," Dino said. "She's a bundle of surprises, wrapped around a stick of dynamite."

"I like that description," Carly said. "Even if it is rampant hyperbole."

"Hyperbole must be near the truth, to be effective," Stone interjected.

"Touché," Carly replied.

After they finished eating, the group lingered over coffee and brandy.

"I have an idea or two," Billy said, lowering his voice.

Dino looked at his watch. "I think that's my exit line," he said. He shook Billy's hand again and departed.

"I've never seen Dino leave early like that," Carly said.

"There are things he doesn't need nor want to know," Stone said.

"Of course," she replied.

"It occurs to me," Billy said, "that I'm the only one in this little cabal that Alexei doesn't know on sight. I think it is to our advantage to preserve that little edge, since it seems to be the only one we've got."

"And you have lots of faces," Stone said. Teddy was a master of disguise.

"I brought a little case with me," he replied. "I'll see what I can cook up that will preserve my unfamiliarity with the little file of faces in Mr. Gromyko's brain."

"Now we have two edges," Stone said, "or as many as you can hatch from your little case."

"You say that he maintains offices in Little Italy?"

"Well, he sometimes does business from a private dining room behind the bar at a restaurant, which is closed at lunchtime," Stone said. "I'm not sure it qualifies as an office, but I doubt he'd be there. The FBI is looking for him, and it's probably one of the first places they checked."

"Forget the office, then. We need only a short time in his presence, long enough to conduct a single transaction and to depart the scene without sprinting into a subway station with his minions in hot pursuit."

"I've heard he sometimes buys fruit on his way home," Carly said, "à la Brando in *The Godfather*."

"I can't establish myself as a fruit seller on the street, without exciting the enmity of half a dozen others who ply the same trade. I would be noticed in a trice and dealt with in the same moment. Besides, if he can't visit his favorite restaurant/office at the moment, I'm confident he won't go anywhere near his home."

"Let's put *The Godfather* out of our minds," Stone said. "Where else does he do business?"

"My notes," Carly said, tapping her temple with a forefinger, "put him constantly on the move, and switching cars frequently, which obviates planting an explosive device in his transportation, à la Trench."

"Quite," Stone said, in an upper-class British accent. "Though doing so would be poetic."

"Perhaps we could find a way to put his paranoia to our own uses," Billy said.

"That's an attractive idea." Stone glanced at his watch. "Why don't you two put some flesh on its bones, and we'll talk again tomorrow. Billy, I'll trust you to get Carly home safely, since you're both staying at my house."

"Of course," Billy replied.

"I'll have Fred drop me off and come back for you in a few minutes." He got up, went outside, and crawled into the rear seat of the SUV.

"Only you, sir?" Fred asked.

"You can take me home and come back for Billy and Carly."

"Yes, sir."

For the rest of the ride, Stone didn't occupy himself with the problem at hand, since he had two devious minds already at work on it.

Gromyko's watchers were gone by the time they reached

home. Fred dropped Stone in the garage, and Stone made his way to the master suite, unbuttoning things along the way. He was sleeping alone tonight, and he needed the rest.

CHAPTER 40

First thing the next morning, Stone sat in his study and waited for his newly formed team of assassins—a film producer and a rookie attorney—to join him.

Eventually, they filed in and sat down.

Stone waited for their plot to unfold.

"We've talked it over at some length," Billy said, "and I've decided that it's best to fall back on old habits and experience. Fewer ways to screw up."

Stone squinted at Billy over his folded fingers and waited. *Old habits?* he asked himself. "What kind of 'old habits'?" he asked Billy.

"I've put out some feelers. Gromyko's been moving around a lot, but I found out where he'll be this afternoon. We wait for him to step out of his nest, then I walk up behind him and put two in his head." As if it were the easiest thing in the world. "It's the easiest thing in the world," Billy said.

"And Carly's role in this?"

"To walk on the other side of the street and observe. It will be good experience for her."

"And how do you get your own ass out of there after shooting the guy in the head?"

"You leave that to me," Billy said, like it was a done deal.

Stone took a deep breath and let it out slowly.

"I know it's not the sort of plan you were thinking about," Billy said. "That's why it will work. It's simple. Nobody will be thinking of it."

"You don't think Gromyko's bodyguard will notice when his charge collapses in the street?"

"Oh, I'm sure he will," Billy said, as if that explained everything.

"Billy, I know your work well enough to know that you will vanish into thin air, but is Carly just supposed to hike up her skirts and sprint away?"

"It's hard to explain, cold, like this," Billy said, "but it will go much better in the viewing."

"This won't be happening in a screening room, but in a public street, likely a crowded one."

"The more witnesses, the better," Billy replied. "They'll each have their own story to tell, and none of them will match."

Stone took another deep breath.

"I want there to be no way for Carly to be associated with what happens; that's the only way I'll feel comfortable with her participation." He packed as much finality into that sentence as he could manage.

"No one will be paying her any kind of attention."

"Is that a promise?"

Billy drew an *X* across his chest with his finger, and said, "Cross my heart."

Stone looked at Carly. "And you're okay with everything?"

Carly shrugged. "Peachy keen. I've heard enough from Billy to believe that he knows what he's doing."

"The first sign you might be in danger, I want you off the street and unobservable from any point of view," Stone demanded.

"I promise," Carly said.

"One more thing," Stone said to Billy. "I'd like a box seat for the performance."

"I'm sorry, Stone," Billy said, "but you're too likely to be spotted as the pain-in-the-ass uptown lawyer you are. Remember, some of those people on the street will have already attended briefings on how to 'rub you out,' as they used to put it. Would a hi-def video feed of the action do it for you?"

"As long as Carly doesn't star in it," Stone said. "I don't want the Greek's buddies to come looking for her."

"Hey," Carly interjected, "me, neither."

"Nor *I*, either," Billy said.

"Sorry. I sometimes forget that Billy is my new high school English teacher."

"We all learn from Billy," Stone said. "When does the curtain go up on this little drama?"

"This afternoon at five, more or less, sharp. I have it

on good authority Gromyko will be having an early dinner only a couple blocks from where he is staying."

Stone threw up his hands. "I surrender, Billy. You always best me."

Billy excused himself and left, but Carly held back for a moment.

"What?" Stone asked.

"Any advice?"

"Any advice I give you would run along the lines of taking the next bus out of town, and I sense that's not what you have in mind."

"Nope," she said, confirming his judgment. Then she was out of there.

Stone sat in his study alone, trying to picture how this thing could work without getting both Billy and Carly killed, instead of the guy who was supposed to get killed. It didn't work.

CHAPTER 41

Carly sat in her bus seat and watched the back of Billy Barnett's head, but not too closely. She didn't want to get caught doing that. The bus stopped, and she got off. Billy stayed with the bus and rode away.

She started up the street in the direction she had been told to walk and looked for a shop. It had been left to her to choose what kind. Krispy Kreme looked pretty good to her. It was busy, but not too busy.

Keeping her back to the street, she entered and waited for the woman ahead of her to conclude her business, which seemed to include feeding a large birthday party. Carly pressed the button in her brain that said, *Calm*, and she instantly was. She read the overhead menu a couple times, then stepped up when the woman left with her purchase.

"Yes, ma'am?" the woman behind the counter said.

"A dozen chocolate glazed and two dozen original glazed," Carly replied.

Her order was filled, and a price mentioned. Carly handed her a fifty.

There was a flash of green being counted, and the jingle of change. Carly dropped the coins into the charity collection jar and stuffed the bills into her pocket. The handle of a shopping bag emblazoned with the product name was thrust at her, and she accepted it and turned toward the street.

As she reached the door and opened it, a Vespa motor scooter flashed past her, and a moment later, two crisp pops sounded. She turned right and began to walk unhurriedly up the crowded street, swinging her shopping bag.

Then there was a kind of collective gasp, and a small girl screamed. People fell away from a man lying facedown in the gutter. Carly stopped and stared at the inert form.

"Somebody call 911," she said to no one in particular. People moved around the man like leaves in a stream around a rock, so she made the call herself, using a throwaway Billy had given her.

"911, what is your emergency."

"It looks like a man got shot in the street," Carly said. She gave the approximate address, but not her name. "He seems to have bullet holes in his head." She hung up, put the phone away, and walked around the seeping form in the gutter.

Half a block behind her there was a low moan, repeated, from a police car, and a crowd began to encircle the man and stop. A police car nosed its way into the circle, and two cops got out. One of them was talking into a handheld radio.

"Man down in the street," he said, then bent and examined the man. "What appears to be a pair of gunshots to the back of the head." He felt the man's neck. "Unresponsive. I can't find a pulse."

The other cop walked in Carly's direction. "Lady, what did you see?"

"What you see now," Carly replied. "That's it."

The cop looked for a more responsive customer, and an old lady accommodated him, talking rapidly.

Carly turned and walked away. She reached into the bag and pulled out a donut and took a big bite. Ahead of her, a woman got out of a cab, and Carly got in.

"The Strand Bookstore," she said and gave the address on Broadway and East Twelfth Street.

Shortly, the cab stopped, Carly got out, leaving the Krispy Kremes behind, and Billy got in, giving the driver an uptown address. Carly walked into the huge bookstore and shopped around, choosing two biographies, Eleanor Roosevelt and Kate Lee. She paid in cash and left the store, having shed her raincoat, and with a new shopping bag.

Stone watched the TV intently and saw Carly go into the Strand, then lost her. Somebody got into her cab and drove away. There was nothing else to see.

Ten minutes later, there was a tap on the rear street entrance to Stone's office, and he let in Billy Barnett, who was, somehow, dressed differently than when he had departed an hour ago.

"Did you see everything?" Billy asked.

"No," Stone said honestly, "just a shot of Carly getting out of a cab at the Strand, and you getting in."

"Then you missed all the action," Billy said. He picked up a remote control and rewound the video, then played it in slow motion.

"I still missed most of it," Stone said.

"Then so will the police," Billy said, holding out a shopping bag. "Krispy Kreme?"

Stone looked into the bag and saw donuts, but didn't take one. "I'm confused," he said.

"And that's a good thing," Billy said.

Shortly, Carly entered the house and flopped down in a chair. "Got another Krispy Kreme?" she asked, and Billy offered her one. She chewed reflectively for a moment. "Well?" she asked nobody in particular.

Everybody stared at Carly.

"Well, what?" Stone asked.

"That's the best question you could have asked," Billy said. He played the recording again, then recordings of two television stations.

"I see nothing," Stone said, "except the back of Carly's head, once or twice."

"And nobody's looking for the back of Carly's head," Billy said.

"It's all too obvious to believe," Stone said.

"Correct."

"And I did what Billy said," Carly remarked. "I learned a lot."

"What did you learn?"

She held up her Krispy Kreme. "I learned that nobody cares about a lady buying donuts. And that, if you just do what you would normally do in the circumstances, you're not a suspect."

"Well," Stone said, "in many years of trying to solve homicides, I didn't look for ordinary people doing what they did. I looked for obvious suspects, and usually, I found them."

"That's because the perpetrators were influenced by their own actions," Billy said. "They were furtive because they knew they were guilty, or were going to be. If you're going to commit a murder, state of mind is everything."

"You didn't explain that before," Stone said.

"No, and that's because an explanation would have altered your state of mind, and you would have been looking for missteps, instead of overlooking them." Billy set the remote control down. "I have a plane to catch, but keep me updated on any developments, Stone. Let's stay in close touch." He nodded at Carly and headed for the door.

Dino sat and watched the videos of the murder from neighborhood security cameras. He thought he caught sight of Carly's head once, but then he lost her. He saw two women carrying shopping bags from Krispy Kreme, but what the hell, it was right there on the corner, and that would have been entirely acceptable and actionless. Who cared who bought donuts? And he would expect that, if questioned, Carly would have had an ironclad alibi: she was buying donuts on a whim.

"Good luck with this one, guys," he said to his team. "Question all the hit men on our list. Nobody's going to miss Gromyko. You'll come up dry." Everybody filed from the squad room and went to work, while Dino returned to One Police Plaza, to issue a statement and return the phone calls of media people.

That night, he slept the sleep of the ignorant.

CHAPTER 42

In the wake of Gromyko's murder, the council meeting was delayed until one AM. Two members of the family's inner circle were in Boston and had to fly back. A third was in the hospital, recovering from a gallbladder operation, and had to be transported to the location via ambulance.

The Bean Counter was in attendance, though word of his earlier meeting with the Greek, and its antagonistic nature, had made the rounds, and few of the others would even make eye contact with him.

The gathering was chaired by Igor Krupin, the senior member of the council and the person who oversaw operations near the docks. "For the third time in less than a year we are meeting because our leader has been murdered. I know you all are as angry as I am. And that—"

"Do we know who did it?" The question came from

Dmitri Asimov, the man who oversaw the family's distribution business.

"I was getting to that."

"Get to it faster."

While the question of who was responsible for the Greek's death needed to be answered, the Bean Counter knew the more important question was who would lead the family now. Because until that was decided, the tension between not just Krupin and Asimov but every man sitting around the table would only escalate.

Krupin glared at Asimov. "At this point, it's too early to know who is behind it."

A chorus of upset voices broke out.

Krupin held up a hand. "Quiet. Quiet!"

"Of course, we know who did it," Asimov said. "Gromyko himself was sure if he were killed, it would be at the hands of the lawyer, Barrington."

"Barrington?" another man at the table asked, incredulous. "The same Barrington who was responsible for the death of the elder Pentkovsky brother?"

"*And* the first Greek."

Everyone turned to Krupin.

"Why hasn't he been killed yet?" a man sitting next to the Bean Counter asked.

"We can't afford to make a mistake in this," Krupin said. "We must make sure before we take action. Trust me when I say I am doing all I can to resolve this quickly."

"You say that like you're in charge," Asimov said. "I don't recall us making that decision."

"Someone needs to—"

More shouting broke out, all of them trying to drown out each other. All but the Bean Counter. Once he felt the cacophony had gone on long enough, he stood and looked around the table, without saying a word.

One by one, the men fell silent.

"Mr. Asimov has a point," the Bean Counter said. "Our first priority should be deciding who will take over the family. Only when we have a new leader can we discuss dealing with whoever is responsible for the Greek's death."

One of the men scoffed. "I suppose you're going to suggest you should be given the job."

The Bean Counter absolutely thought he was best suited, but he was smart enough to know now was not the time. "You would be wrong. I do have a suggestion, but it is not me."

"Who?" Asimov asked.

"There is really only one choice," the Bean Counter said, then smiled.

CHAPTER 43

Stone decided to spend the next day at home. Given the Greek's death, it seemed a reasonable precaution.

He had suggested the same to Carly.

"Why should I worry? I didn't do anything," she said.

"You did help plan his execution."

"They don't know that, do they?"

"Probably not, but look at it this way. You are a known associate of mine, and I am someone Gromyko very much wanted dead. If they can't get to me, they may try to get to my friends. Have you forgotten they already kidnapped you once?"

"That's one more thing I'll never forget."

"I thought not," Stone said. "If you stay here, you are under the protection of Mike Freeman's people, the best in the business."

"I suddenly have the desire to work from home today."

"I thought you might."

Stone had just finished a midafternoon video conference when Dino called.

"Everyone still breathing over there?"

"Last I checked."

"That's good to hear. Dinner?"

"Always."

"Clarke's or Patroon?"

"Here, I think. No sense in tempting fate."

"Works for me. What time should we be there?"

"Six-thirty for drinks?"

"See you then."

Stone handed drinks to Dino and Viv as soon as they walked in, and then refreshed the one Carly had been drinking.

"Did you see the news?" Dino asked.

"That depends on what news you mean," Stone said.

"About Alexei Gromyko."

"If you're about to tell me he's dead, then yes. I heard something about that."

"I'm sure you did. I was talking about his funeral."

"That I hadn't heard. What about it?"

"It's tomorrow."

"That's fast," Carly said.

"It is," Stone said. "I guess there's no question about the cause of death."

"As the coroner said to me," Dino said, "'the means of Mr. Gromyko's demise is self-evident.'"

Carly frowned. "That still doesn't explain why they'd bury him so soon."

"I can think of a reason," Viv said. "A good one, too."

"I'm all ears."

"So they can get it over with."

"That isn't very respectful."

"Maybe not," Dino said, "but Viv's right."

"I never get tired of hearing you say that," Viv said.

"I've noticed."

"I still don't get it," Carly said. "Gromyko was the head of the family. Shouldn't his funeral be a big deal?"

"From what I heard, it will be," Dino said. "We're calling in an extra one hundred fifty officers to handle security."

"The police doing security for the mob." Carly shook her head. "That has to be a first."

"Far from it," Stone said. "Do you think there's never been a funeral for the head of a crime family here before? The last thing anyone wants is a shootout taking place at a cemetery."

"It would be convenient," Carly said, seriously.

"There are times when I worry about what's going on in that head of yours," Stone said.

"Did I say something wrong?"

"I guess that depends on your point of view."

She thought for a moment, then said, "No one has told me yet why it's happening so fast."

"Simple," Viv said. "Everyone's focus will be on the

funeral. Until it's over, everything else they had going on will be put on hold."

"Like day-to-day graft," Stone offered.

"Or dealing with the person who had the head of their organization killed," Dino said.

Stone blanched. "I could have gone without hearing that."

Helene stepped into the doorway. "Dinner is ready."

"Great, I'm starved," Dino said.

"Me, too," Carly said.

Stone, however, had lost his appetite.

After a meal of grilled salmon, baby potatoes, and asparagus—most of which Stone just pushed around his plate—they retired to the study for a glass of port.

"You look like you have something on your mind," Carly said to Stone.

"What gave you that idea?" Dino asked. "The fact that he barely ate anything? Or that he has yet to take a sip of his drink?"

"Both."

"It was the drink for me," Viv said.

"Me, too," Dino agreed.

"That obvious, am I?" Stone said. "I have indeed been thinking. Perhaps a trip to Maine is in order."

"The last time I was there, people were shooting at us," Carly said.

"The idea would be that going there would prevent that from happening again, here."

"When are you thinking of leaving?" Dino asked.

"I have a meeting tomorrow morning I can't get out of, but I'd like to be on my way by the afternoon."

"Making yourself scarce right after the funeral is not a bad idea," Viv said.

"I think you all should come with me."

"The whole if-you're-not-around-they'll-come-after-your-friends thing?" Carly asked.

"In a nutshell."

"I'm in. I can work remotely this week."

"I'll have to rearrange a few things," Viv said, "but it's doable."

"Then count us in, too," Dino said.

"Great," Stone said. "Let's plan for wheels up from Teterboro at one PM."

CHAPTER 44

Peter Greco sat in a row of chairs in a thick forest of headstones in a Queens cemetery and listened to a priest drone on in Russian, which he understood only poorly. The Greek's wife, Olga, who was twenty-odd years the junior of her late husband, sat at Peter's elbow and made snuffling noises while clutching his arm. Peter was conscious of the breast pressed against him and of her cleavage, which looked good in black.

He had not planned on attending the Greek's funeral. He had assumed he and his family would have been preparing to move across the country, where they would be under the protection of the FBI. But yesterday he had received a call from the Bean Counter, who had asked him as a personal favor to attend and had guaranteed his safety. So, with some reluctance, he had come.

The service ended, and Olga turned to him. "I would

like you to come to my house for a glass of tea," she said, "and there are those of our community who wish to speak to you."

"I've already attended one funeral this week," he said. "I would rather not star in another."

"No one wants you dead. Quite the opposite."

Intrigued, he followed her to a black limousine and took note of her shapely buttocks as she bent to enter the car. When he seated himself, he was surprised that she occupied her seat in such a way as to keep herself thigh-to-thigh with him.

"Who wishes to speak to me?" he asked her.

"People," she replied, then spoke no more for the remainder of the ride. They entered the old, but well-kept house, and she pointed to the dining room door. "In there," she said. "I will wait for you upstairs."

Peter opened the door and peered into the room. A group of men rose as he entered, then settled themselves in the chairs around the table.

"Shall we speak Russian?" an elderly man asked him.

"Please, no. I haven't spoken it since I was very young. Just English." What the hell was this about?

"Very well," the man said. "We are here to remind you of your duties."

"My duties?" Peter asked. "What duties?"

"The first is revenge," the man said, and there was a positive rumble from the group.

"Revenge toward whom?" Peter asked.

"Toward the man who killed the Greek, or rather, who ordered his death."

"And who might that be?"

"This person, Barrington, the lawyer."

"And how did you come by this information?"

"Alexei predicted his death by Barrington's hand or on his order."

"That's not evidence enough to suspect Barrington in his death," Peter said.

"Is the Greek's word not sufficient for you?"

"No," Peter said, "but I don't see why it even matters what I think."

"It matters," the man said, "because as our new leader, you must take charge of the effort against Barrington."

"Leader of what?" Peter asked, baffled.

"Why of our family and its businesses," the man said. "The Gromykos did not work out. It's time for a Pentkovsky to lead the family again."

"I cannot accept that," Peter protested. "I had already expressed to the Greek my intention of leaving the group and governing my own existence. That has not changed. Besides, I do not believe Barrington capable of ordering a murder."

"You are naive," the man said.

"Perhaps so. But I am Peter Greco now, not Egon Pentkovsky. And I must decline any participation in the family's affairs, and certainly anything to do with revenge against someone who has been wrongly accused."

"We shall see," the man said.

"We shall *not* see," Peter said. "And I will not be pressed into replacing the Greek in the family, whose activities I wish to have nothing to do with."

The man sighed, ignoring Peter's protests. "And then there is the matter of the Greek's widow," he said.

"I'm sure he has provided for her generously."

"He has provided money and the house," the man said, "but those are not all the woman needs. She is young and beautiful, and she must be attended to."

"I have a wife and family, and I attend to those," Peter said. "She will have to look elsewhere for attention."

"Then you must explain that to her personally."

"Please explain to her on my behalf."

"She is waiting upstairs to speak to you of this. It is ceremonial, on such an occasion. Go to her now. You will not be disturbed."

Peter stood up. "Gentlemen, I thank you for your interest in me, but I have quite another life to live, so I must leave you now."

"As you wish," the man said. "We will speak further."

Peter left the room, closing the door behind him. As he walked toward the front door, wondering how he was going to transport himself back to the city, he heard a voice from upstairs.

"Peter?" Olga called.

He stopped. "Yes?"

"Please come upstairs."

He was going to have to explain his position to her and be done with this foolishness. He climbed the stairs and came to a half-opened door.

"Please come in," she said.

Peter pushed the door further open and entered the room. The door closed behind him.

"I knew you would come," she said. "I knew it in the car."

He turned and saw her standing next to the door, and she was entirely naked.

She quickly closed the distance between them and pressed her body against his. Peter instinctively put his arms around her in a gesture of comfort, but what he felt against him had nothing to do with comfort, and everything to do with lust.

Peter's relationship with his wife had cooled over the years, and his response was nearly instantaneous. She did something with his belt buckle, and his trousers dropped around his ankles.

"Come," she said, taking his penis in her hand and leading him toward the bed. His choices were to follow her or to trip over his trousers and fall flat on his face. Before they reached the bed, she had stripped him to the skin, and he fell on the mattress with Olga on top of him. From there, he had only to follow her moves, which he did automatically. She sat on top of him and enveloped him with her thighs.

She was an expert. Again and again she brought him to a near climax, then slowed, until finally, he could hold back no longer, and they climaxed together.

"There," she said, "that's so much better, isn't it?"

Finally, they dressed, and she led him to the limousine parked outside, waved goodbye, and then returned alone to the house. One of the men who had spoken earlier at the family meeting got out of the car and shook Peter's hand.

"Now she is yours to do with as you will," he said. The man embraced him. "And you belong to your family again."

Before Peter could manage a response, he found himself inside the car and it was driving away. He was in a daze and didn't much care where he was going. Eventually, the car pulled up outside an elegant apartment building where he maintained a comfortable pied-à-terre that he had been steering clear of while hiding from the Greek.

The driver held the door for him and, as he got out, pressed a card into his hand. "This automobile is now yours and is at your disposal. I am Boris, and I am yours, too. You may call at any hour." He pressed a red iPhone into Peter's hand. "Use this to conduct business. I will always be nearby. There are two airplanes at Teterboro that are at your disposal as well, one for long flights and a smaller one that can be landed at short fields. I can arrange any flight."

Peter walked into his building and was saluted by the doorman. He took the elevator upstairs and entered the apartment. He was greeted by a large vase of calla lilies on the hall table, and a card read: *From your family. We request a meeting Monday, at noon, unless you postpone. You choose the location. Our number is in your iPhone. The sum of one million dollars has been deposited in your account at the corner bank. You must report it as income on your tax return.*

He went to the bar and poured himself a neat Scotch,

then sat down in his study and sipped it, reviewing the events of the day. Olga, alone, had not sealed the deal, but she had helped. The car, the two airplanes, and the money had done the rest.

The phone rang. "Yes?"

"It's Marla, sweetheart," his wife said. She and his daughters were at the home on Islesboro that Stone had helped them find. "When may we expect you?"

Peter looked at his watch. "Meet me at the airfield at four PM," he said. "How is the house?"

"Just wonderful. Stone Barrington sent some lovely yellow roses. See you at four."

Peter hung up and sipped his Scotch while he thought ahead. If he was going to be in charge of the family, then he was going to do it his way. He would legitimize the family business; it was the only way he could live with his new circumstances. Then there was the business with Barrington. He would have to find a way to deal with that. And finally, there was the testimony he'd given to the FBI. He would call Assistant Director Kinder, see if he could negotiate a deal for time to get the family on the right track.

CHAPTER 45

Peter called Boris.

"Yes, Mr. Greco?"

"What are the two airplanes at my disposal?"

"A Pilatus and a Dassault."

"I want to be flown to Islesboro, Maine, in the Pilatus, to land a four PM. The runway is 2,450 feet."

"I'll call you back." Boris hung up. A few minutes later, he called back.

"Yes?"

"We must leave the apartment at two-fifteen. Your flight will depart at three PM, and you will land at four."

"See you at two-fifteen." Peter hung up, then called Stone Barrington's cell.

"Yes?"

"It's Peter."

"Hello, are you in the new house?"

"My wife is, and she thanks you for the roses. I should arrive at the airstrip at four. Are you in New York?"

"Just departing Teterboro. I'll be at Islesboro myself in an hour."

"May we meet at the new house at, say, five? I have business to discuss with you."

"All right," Stone said. "I was surprised to see you on a report on the morning shows about the Greek's funeral."

"Not as surprised as I was to be asked to attend. Things are . . . never mind. We can talk more about that later. See you at five."

Peter sat in the copilot's seat in the Pilatus. It was an airplane he knew but had not flown in: a single-engine turboprop, good on short runways. Noisy, without a headset.

"Islesboro dead ahead," the pilot said.

Peter checked his watch. Five minutes till four. "Good." He watched closely as the pilot set up his approach: flaps down, throttle retarded. The pilot aimed at a spot just below the runway numbers, then set the airplane down, reversed the prop, and braked. The airplane landed with a thousand feet to spare.

"Excellent," he said to the pilot, then he got up, went to the door, and walked down the airstairs. Marla waited in a Mercedes station wagon.

"Whose airplane?" she asked, after a kiss.

"Ours, when we want it," he replied.

"Ours?"

"I'll explain it to you later. Stone Barrington is coming for a drink at five o'clock. After, he and I will need to talk privately for a while. I'll ask you to excuse us."

"All right. How did it go with the family?"

"Better than expected. I'll explain that, too."

She nodded and drove on, chattering all the while. She unpacked for him, then came down in time to greet Stone.

Stone accepted a Knob Creek and looked around. "Looks like people live here," he said.

"We can thank Jimmy Hotchkiss's wife for that," Peter said. Tracey Hotchkiss was an expert interior designer. "Marla, will you excuse us for a while?"

"Of course," she said, and took her drink with her.

The two men settled themselves in comfortable chairs before a small fire. "My situation has changed."

"How so?"

"Today, my family chose me to succeed Alexei Gromyko. It was against my wishes, but they prevailed."

"You understand that I cannot discuss matters of a criminal nature with you?"

"That won't be necessary. I just want to tell you that I am, starting on Monday, going to decriminalize all my family's business interests."

"From what little I've heard of the extent of your businesses, that will be quite a task."

"Not as great as you may think. Gromyko, during his tenure and under the influence of the Bean Counter, began structuring the majority of the businesses involved into corporate groups or holding companies that are transparent and pay taxes. There are, of course, a few enterprises the family has been involved in that can't be incorporated for legal reasons. While Gromyko had planned to keep those businesses, I will be eliminating them completely."

"A good start. What about your testimony to Assistant Director Kinder?"

Greco grimaced. "That is a delicate matter. I talked to Kinder before leaving the city and let him know about the change in my status, and that I would no longer be cooperating with his investigation."

"I imagine that didn't go over well."

"I explained to him what my plan was and asked that he give me a little time before he pursued the family further."

"And he agreed to that?"

"He said he would take it under consideration."

"That's more than I would have expected," Stone said.

"I believe my frankness in our discussions at your house created some goodwill between us."

"That may be so, but don't expect it to last forever."

"I asked him to give me two months. I guess we'll see if he will."

"So, are you Egon Pentkovsky again? Or is there a new name I should call you?"

"I'm keeping Peter Greco for now. I know my Pentkovsky roots are the reason they gave me the job, and I will embrace that. But I also feel going by Greco will be a fresh start for everyone." He leaned forward. "Stone, I want you to know that I attach no blame to you in the matter of either of the Greeks' death, nor those of my brothers."

"I'm glad to hear it," Stone said. "Does that go for all the members of your family?"

"I have not spoken to all of them yet, but I will see that the word goes out that neither you nor yours be touched."

"I hope the word spreads quickly."

"It will, but do be careful for a few days."

"Thank you for the warning. But having to watch my back touches on what seems to me will be your biggest problem. By that I mean, restraining the criminal activities of the people who conduct the family's operations."

"I will begin addressing that problem on Monday, when I meet with the council. May I retain you and your firm to help with the legitimization?"

"No, because that would entail being involved with criminal matters until you have made everything squeaky clean. My firm, Woodman & Weld, operates at the highest level of the legitimate practice of the law, and they will not be drawn into enterprises that don't meet their ethical standards."

"Understood. May I seek, personally, your counsel from time to time?"

"I'm sorry, no. And we can't meet again until you have completed your cleansing. I will give you this advice, though: the faster you legitimize the better. It is only when law enforcement agencies such as the FBI begin to fail to make cases against you that they will start to back off a bit."

"Thank you for that. And I understand your position. I will not violate your boundaries."

Stone stood up. "Well, then. You might begin your cleansing on this island. Never give anyone here anything to gossip about."

"Good advice," Peter said, and escorted him to his car.

Back in the house, Peter sat Marla down and explained to her what had happened with the family and what her obligations would be, beginning with using only legitimate credit cards for her personal and household shopping. He did not raise the subject of Olga.

"You're really serious about this, aren't you?" she asked.

"Yes, and you must be, too. You can begin with the occasional generous, but not outrageous, donations to local charities. It will be good for people to think of us as donors, when the subject comes up."

"I understand," she said, then came and sat in his lap, surprising him. He found that he enjoyed being surprised.

CHAPTER 46

Stone returned to his own house, where he found his guests—the Bacchettis and Carly—settled in the living room before a big fire. "It's starting to rain," he said to nobody in particular.

"Raining in Maine?" Dino said. "No kidding?"

"We're not in Scotland, Dino," Stone replied. "It doesn't rain constantly here. There are often sunny days."

"Whatever you say," Dino replied, gazing at the rain beating against the windows on the seaward side of the house.

"It's nice weather for the fire," Carly pointed out.

"Thank you for that," Stone said. He poured himself a drink and sat down. "I have news from my meeting with Peter Greco."

"Did he shoot you twice in the head?" Dino asked.

"You will note that my head has only the usual number of holes."

"Did he miss?"

"This morning, after the Greek's funeral, Greco was appointed the new head of the family."

"You should be wearing full body armor," Dino counseled.

"On the contrary, he is meeting with his council on Monday, and at that time he will give them instructions that neither I nor mine—which includes you lot—are to be touched."

"I'm happy to be included in your lot," Dino said.

"We are still to be careful until word has filtered down to the soldier level."

"Careful?" Carly asked. "What does that mean?"

"Don't go out alone," Dino said, "and when you do, go armed."

"That's what we were doing anyway, wasn't it?" she asked.

"Then just keep on doing it. Peter says he's taking all the family businesses fully legal."

"Does that include the murder and mayhem part?" Viv asked.

"Apparently, Gromyko incorporated most everything and started paying taxes."

"You can do that with murder and mayhem?"

"Businesses have been doing that for centuries. But I do believe he intends to divest himself of that particular avenue."

"Will wonders never cease?" Viv asked.

"Wonders remain to be seen," Stone replied. "I will say that I am encouraged by what he told me."

"You are easily encouraged," Dino said.

"Perhaps, but I am not credulous. I've gotten to know him a bit, and I appreciate what he's trying to do."

"Does that mean you're his lawyer now?"

"No, it doesn't. I've explained to Peter that Woodman & Weld deals only with upright clients."

"Some of your clients will be surprised to hear that," Dino said.

"You won't find a cleaner, more ethical law firm in the city," Stone replied.

"Okay," Dino said. "I suppose I can stretch your credibility that far."

"I'm so grateful," Stone said.

"Stretching calls for another drink," Dino said, waving his glass at Stone.

Stone complied, and Dino sipped.

"What about the meeting with the FBI?" Carly asked. "Did I remember all that for no reason?"

"Your memory was used as intended. What Peter told the FBI didn't go away just because he switched sides. As for his relationship with the feds, he has requested a grace period from Kinder, to clean things up."

"And Assistant Director Kinder went for this?" Viv asked.

"That remains to be seen."

"My money is on a big fat no," Dino said.

"My money is staying firmly in the bank," Stone said.

Dino took a drink, then said, "So, what are we going

to do with ourselves on this little vacation, besides watch our asses?"

"What you usually do," Stone said. "Sail or go motorboating, shoot skeet, fish, swim in the sea—like that."

"At least half of those activities are precluded by the local water temperature," Dino said.

"Get used to it. It's not going to change."

"Something occurs to me concerning safety," Viv said.

"What is it?" Stone asked.

"You better let Ed Rawls know about this semi-truce you have with Peter G. before he shoots the man's eye out."

The doorbell rang.

"I'll bet that's Ed," Dino said. "The scent of single-malt Scotch has reached his house."

Stone peered through the peephole, then opened the door. "Speak of the devil."

"The devil drinks brown whisky," Rawls said, handing Stone his shotgun, as another man might hand him his umbrella, which he was also carrying and handed over. "So, you're talking about me already?"

"We couldn't help ourselves," Dino said.

"Take a seat and be briefed," Stone said. Rawls did so.

"Is the war over?" Rawls asked.

"How prescient of you to ask," Stone said. "Peter G. has assumed Gromyko's former position. He is calling off the mob's dogs and taking his operations legit."

"Well, shit," Rawls said. "How am I going to entertain myself?"

"Just keep on doing what you're doing, until we get the all-clear."

"You mean keep shooting Russians?"

"No, unless violently provoked."

Ed took a big swig of his Scotch. "No fun at all, eh?"

As the group chatted on, Stone got up, retrieved an umbrella from the holder near the front door and exited the house onto the back porch. He took a seat in a damp, Weatherend love seat, and was soon joined by Carly.

"Hey, there," she said.

"Haven't I seen you someplace before?"

"I'm glad I'm, at least, a familiar face."

"You're more than that." Stone put an arm around her.

She snuggled closer and gave him a little kiss on the neck.

"What was that?"

"That was a little kiss on the neck. Would you like a bigger one, perhaps somewhere else?"

He looked down at her and got a kiss full on the lips, with a little tongue thrown in.

"Hang on," he said, feeling a stirring.

"Onto what?" she asked, placing a hand on his thigh.

"Didn't I tell you not to get entangled with anyone at the firm?"

"No. You told me not to let Herb Fisher get me into bed. You said nothing about you."

"Was that what I said?"

"If you want the exact quote, I can give it to you."

"I'm sure you can."

She lifted her hand. "Are you saying that rule applies to *all* the lawyers at Woodman & Weld?"

Stone floundered, trying to come up with an answer.

"I mean, you didn't bring a girl besides me on this trip. I'm inclined to take that as inviting." Her hand settled back on his thigh. "Am I wrong?"

"I'm not sure there's a right answer to that question."

"Of course, there is," she said. "Tell you what, you think about that for a while, and I'll wait until you're in bed for a straight answer, so to speak."

She kissed him again, even better this time, then got up and went back into the house, leaving him to talk his anatomy down to a non-visible size. He picked up his umbrella, opened it, and walked out into the rain. The wind had dropped, but the rain continued. He kept his back to the windows, until normalcy had returned, then strolled down to the dock and inspected the boats—the Hinckley picnic boat and the Concordia sailing yacht.

Nobody took a shot at him, so he went back into the house. Carly was nowhere in sight.

"Wet enough for you out there?" Viv asked.

Stone shook like a dog. "I think I'm going to need to change clothes."

"Well, Dino and I are going to repair to the guesthouse for a little nap before dinner," she said, ogling Dino into action.

"Enjoy," Stone said.

"You, too," Viv said.

"Wait. Where's Ed?"

"I told him we wouldn't be eating until eight-thirty

tonight," Dino said. "So he went home for a bit and said he'd be back in time to eat." He and Viv left.

Stone put away his umbrella and headed upstairs to the master bedroom. He walked into the room and found Carly in his bed, apparently naked, but with a sheet covering her breasts. She was reading an old *New Yorker* by a bedside lamp.

She reached up and switched it off. "Hello again. It took you a little longer than I thought."

Stone got out of his wet clothes, and she held the sheet up to make room for him, and to give him a better view of her body, the beauty of which exceeded his suppositions.

"I have a feeling I'm going to regret this," he said, kissing her on a nipple.

"Not if I have anything to say about it," she said, reaching for him.

"That speaks volumes," Stone said, giving himself to the moment.

A little later, Stone lay on his back, taking deep breaths.

"Are you going to need resuscitation?" Carly asked.

"Whatever that was worked just fine."

Carly laughed, something she did not do a lot of. "You can have a rest, and we can resume after dinner."

"No need to wait," he said, turning toward her. "It's my turn, anyway."

"You won't get an argument from me," she said, running her fingers through his hair, steering him home.

CHAPTER 47

Stone woke to find Carly gone from the bed, then she emerged from the bathroom, toweling herself.

"Awake at last!" she said.

"Sort of," he replied, kicking off the covers.

"I like the two-of-everything idea," she said, indicating the baths and dressing rooms. "No waiting."

Stone shaved and showered. And when he had dried his hair and toweled off, he came back to the bedroom and found Carly gone, and her clothes in the female's dressing room.

So much for staking out her territory, he thought. He changed into fresh khakis and a clean shirt and slipped into a pair of alligator loafers and a blue blazer, with New York Yacht Club brass buttons, then went downstairs. Everyone was there, except Carly.

"Where's the girl?" Dino asked.

"Beats me," Stone replied, and it did. "Kidnapped, perhaps?"

"I doubt it," Viv said.

Stone fixed himself a drink. "Anybody for a refill?" he asked the room.

"We're good," Dino said.

Stone sat down and switched on the evening news.

"Reputed Russian mob boss Alexei Gromyko was laid to rest today after being shot on the street in broad daylight," the news reader said. The camera panned the group at graveside from a distance.

"Hey," Dino remarked. "Who's the babe sitting next to Peter?"

"That must be the widow," Stone replied. "Though I've never seen a widow at graveside sporting that sort of cleavage."

The TV moved on to the next story.

Dino answered a knock at the door and let Ed Rawls in. "You know where the bar is."

"I have a special, inbuilt range finder that seeks out the bar in every location," Rawls said, helping himself to the Scotch.

Carly appeared at the top of the stairs and made her way to Rawls's side. "Me, too, Ed."

Ed handed her the drink he had poured for himself, then started over.

"Was that really Gromyko's widow on TV?" she asked.

"As far as we know," Stone replied.

Stone switched to CNN and found the same graveside cleavage staring at him.

"She's an all-star," Viv said.

Mary, the housekeeper, rescued them from the news. "Dinner's on," she said.

"That would be lobster," Stone said. "It's always lobster the first night back." He found himself seated next to Carly, with Dino on her other side.

"Did you have a refreshing nap, Carly?" Viv asked sweetly.

"Indeed, I did," Carly said. "The rain does it for me every time."

"You know, one of my detectives told me something I didn't know about Russian crime families," Dino said.

"That would be just about everything," Viv said.

"He said that a childless young Russian widow becomes the ward, so to speak, of the succeeding family head, and that he is obligated to give her a child."

"Well," said Viv, "that's useless information, if I ever heard it."

Dino jerked a thumb at her. "She says that about every bit of news I bring home."

"I don't argue sports with you," Viv said. "Unless Mrs. Gromyko qualifies as a sport."

Ed Rawls spoke up. "I got a glimpse of Greco's wife on the island today."

"How's her cleavage?" Dino asked.

"Oh, Dino," Viv said, "shut up."

"That's my cue to shut up," Dino said, and did.

On his way back to his place after dinner, Ed Rawls's phone rang with an unfamiliar number. He let it go to voicemail and then played back the message.

"It's Peter Greco. If you could, please call me back."

Curious, he hit redial.

"Hello?" The male voice on the other end was the same voice that had left the message.

"Mr. Greco."

"Mr. Rawls, thank you for returning my call. I was wondering if you might have time to discuss a job."

"Sorry, not interested."

"Could we at least meet so that I could explain the parameters?"

"I don't work for the mafia."

"This would be both for and against. All I'm asking is that you hear me out."

Despite himself, Ed was interested. "Tomorrow morning, ten AM. Grindle Point Lighthouse."

"See you then."

CHAPTER 48

Stone woke at eight AM, but thanks to Carly, he did not get out of bed until after ten.

"Join me for a shower?" she asked as she crossed the room.

"Why do I get the feeling there won't be much showering?"

"I promise, every inch of you will be clean by the time we finish."

"How can I say no?"

Forty minutes later, Stone was dressed and feeling cleaner than he could ever remember.

"Shall we see about getting something to eat?"

"Please," Carly said. "I'm starving."

Dino was reading the *Times* in the living room when they came downstairs. He folded the paper dramatically in his lap upon seeing them.

"Viv had to stop me from sending out an APB when you didn't come down for breakfast," he said.

"A little out of your jurisdiction, isn't it?" Stone said.

"You forget, I happen to know the local constabulary. One call and the island's on lockdown."

"Where's Viv?" Carly asked.

"Out for a walk."

"By herself?" Stone asked, concerned.

"She's armed with one of your shotguns, and two of Strategic Services' best are trailing her."

"Glad to hear it."

"Do you really think the Russians will try something here?" Carly asked.

"I'm hoping they don't even know where we are," Stone said. "But no need to tempt fate."

"I suppose."

"Come on. Let's see if Mary can whip us up something to eat."

"Don't eat too much," Dino called after them. "Viv wants to have lunch at the golf club and play a round."

"You hate golfing," Stone said.

"Not as much as I love my wife."

They had lunch at the club at one PM.

"This is magnificent," Carly said, after taking a bite of her stuffed pork chop.

"They have a new clubhouse chef," Stone said.

"I'd say he's a keeper."

"I couldn't agree more." Stone's chicken cordon bleu and sauteed mushrooms were particularly tasty.

"Uh-oh," Viv said, looking out the windows.

The others followed her gaze to the patio outside, where the first splatters of rain had begun dotting the ground.

"Maybe it's only a sprinkle," Carly said.

The words were barely out of her mouth when the sputtering shower turned into a hard downpour.

"I guess golf is off the agenda," Dino said.

Viv turned the evil eye on him "You don't have to sound so happy about it."

"Happy? My heart is breaking."

"In glee," Stone said.

"That's not helping," Dino said. "Unless your goal is for me to sleep on the couch in your living room."

"I retract my previous statement."

"I thought so."

"As clumsy as his delivery was," Viv said, "I fear Dino is right. Even if the storm passes, who wants to trudge around on wet fairways?"

Carly raised her hand.

"Of course, you do," Stone said. He gently lowered her arm. "Lucky for you, you have mature friends who can save you from making that mistake."

"If you say so."

"I do."

They replaced the round of golf with drinks in the bar, where they could monitor exactly how wet the course was getting.

They had just ordered their second round when Ed Rawls walked into the room.

"Ed!" Carly said.

"Join us," Stone said.

They made room for him at their table.

Dino motioned for the server to come back, and said to Ed, "What are you drinking?"

"A Scotch, please. Laphroaig. And make it a double. It'll be my last for a couple days."

Dino placed the order.

"Last?" Stone said. "Don't tell me you're swearing it off?"

"Blasphemy," Ed said. "I just need to keep sharp."

"Something we should know about?"

"Probably. But it might be better if you didn't know the details. I'll just say this. It's best if you stick to your place on Monday. No wandering around."

"What's happening on Monday?" Dino asked.

Carly looked at Stone. "You said that's when Greco's having his meeting."

"I did." Stone had made the same connection. To Ed, he said, "Are we to assume said meeting is taking place here on Islesboro?"

"If it was, staying close to home would be a good idea," Ed said. The server arrived with his Scotch. Ed thanked him and took a healthy sip.

"And you know this because?" Viv asked.

Dino raised both of his hands. "Wait. I'm not sure I should be around to hear the answer to that." He started to stand, but Ed motioned him back down.

"I never said I knew anything."

"Greco must have told him," Carly said. "Unless you did, Stone."

"I had no idea where the meeting was going to take place," Stone said.

"Then that leaves Greco," she said. "Ed, are you doing something for him?"

Ed's brow tensed.

"Carly," Stone said, "that's one of those questions you shouldn't ask."

"Why?"

"You remember when Ed said it would be better for us not to know the details?"

"Yes. Oh, I got it."

"I knew you would."

CHAPTER 49

The group of Russians met on Islesboro at mid-morning Monday, after sleeping on the mainland the night before, then chartering their own ferry. Peter received them on the front porch at noon, and after a glass of something, they sat down to a buffet lunch. When that had been consumed, and the house had been cleared of caterers and his wife and children, he sat down at his dining table with the family's council.

"Once again, welcome to Islesboro," he said. "It is my understanding that I have been chosen to lead the family. So you all say?" The group made affirmative noises. "Very well, I accept, and I will run things pretty much as the late Greek did, absent the murders."

This news was greeted by a cold silence. "Anyone with objections to this policy will be murdered," he said, then paused for them to realize he was joking and laugh a bit.

"I've spent the weekend catching up with current operations and the books. As you know Alexei initiated the policy of conducting ourselves as legitimate businesses, and that conversion will continue apace. As time passes you will begin to know the advantages of legitimacy, particularly of paying taxes. We will gain more from paying than it costs us.

"I want you all to understand that anyone who violates the legitimacy rule will pay dearly. And murder and other violent methods will no longer be practiced." Their expressions remained hostile.

"We will begin with the matter of Stone Barrington," Peter said. "Neither Barrington nor his friends and associates in and outside his law firm will be touched."

"But we have had multiple generations of leadership with his death as an objective."

"All that is over," Peter said.

"What if he provokes us?" someone asked.

"We will not be provoked," he replied. "If Barrington needs to be spoken to, I shall do the speaking. Does everyone understand this rule?" He looked around the table and waited until he got a nod from every man present before continuing. "Good. It is my wish that everyone at this table reach their own home safely tonight." He waited for that to sink in.

"Divisions of responsibility for management will continue as envisioned by Alexei. If changes need to be made, I will make them. While the business will be legitimate, it will not be a democracy, and votes on any subject will not be taken. My orders will suffice. Only those who accept these conditions will be welcomed at this table when next we meet. Others may take retirement as outlined in Alexei's plan. You are all wealthy men and will take salaries from the individual companies. I will award such bonuses as I feel are good for the whole business. It will be expected of all of you to live on the salaries assigned to you. Displays of personal wealth are to be discouraged." Peter continued for another half hour, then closed the meeting and thanked all for their attendance. He wondered how many he would have to kill before they took him seriously.

Ed Rawls, sitting in his living room, took off his headset, and switched off the receiver of the microphones he had planted in the house before the meeting. He called Stone Barrington on his secure cell phone and told him what had transpired. "I'll bet you the son of a bitch doesn't live for another week," he said.

"Don't underestimate him," Stone replied.

Ed decided to tell him what he hadn't at the golf club bar.

"I had a feeling it was something like that," Stone said. "At least, he's hiring the best."

"I'm still not sure if I should have said yes."

"Is he paying you well?"

"He is."

"Then take his money and consider it island pest control."

"Is that your professional opinion?"

"Professionally, I have no opinion on this because this conversation never happened."

"Like so many of our conversations."

"Per your suggestion, we're having dinner here tonight," Stone said. "If you're free, join us."

"I'll check my schedule."

CHAPTER 50

Ed Rawls watched the departure of the Russians from the dock next door and did a head count: two missing. He sent Peter Greco a text with this information and confirmed that the two were the same pair Greco had suspected would be the problem.

In short order he received a reply: **This should be a straightforward assassination. You may shoot to kill. Don't forget the cleanup. Payment will be as agreed and placed in your mailbox.**

Consider it done, Ed wrote back. He walked over to the newly renovated house next door, which he knew to be unoccupied, carrying only a key and a briefcase.

He assembled the Czech sniper's rifle, set up a tripod on an upper deck where it could not be seen from the road, and checked the view from the deck, past his house and dock. As the sun set, a dinghy appeared at the dock of the new house, and two men carrying duffels alit on the dock and tied up the dinghy.

Using a 24x monoscope, he followed them up the dock and past his perch, watching as they waited behind shrubbery for the traffic to clear, then crossed the road. Ed went back to his silenced rifle and began to follow their progress with the rifle's scope. A message came on his radio, through the earpiece. "Anytime now."

Ed aimed at the second of the two figures sneaking up to the house, and sighted him in. He squeezed off a round, catching the man above the ear and felling him. His companion turned around to discover the source of the sound he had heard, and Ed shot him in the forehead before he could collect himself.

Ed spoke into the radio. "All present, dead, and accounted for," he said.

"Excellent," came Greco's reply. "When will disposal take place?"

"Your part, as soon as the sun goes down," Rawls said.

"They will be bundled and delivered to you in the craft they arrived on, at that time," Greco said.

"Good. I will complete my part after dinner, pending receipt of payment."

"Done."

Rawls disassembled and packed the rifle. He walked back to his house, put the case in his safe, then got into his car for the drive to the Barrington house. On the way, he stopped at the mailbox to collect the waiting envelope. When he reached his destination, he took a moment to do a fast count of the envelope's contents, pocketed it, and walked up to the house.

Carly met him at the front door, while the others were just gathering at the table. "Come in, Ed, and take a seat."

Stone looked at him questioningly for a moment. Rawls nodded.

"Welcome, all," Stone said, raising a glass of a fine California Chardonnay to them in greeting.

The others joined the toast, then they all took a seat.

Mary began distributing plates, and Seth, her husband, brought in a large serving dish, set it on a trivet at the center of the table, and removed the lid. Stone was handed each plate and, in turn, filled them with roast pork and apples and sent them around the table.

"Bon appétit," he said, raising his replenished glass. They responded, then dug in.

After dinner and an hour of drinking port and eating Stilton before the fireplace, Rawls excused himself. "Pardon me for leaving a little early, but I forgot to set out the trash."

He drove back to his house, using his iPhone to work his way through the security system, then he walked down to the dock. Tied up on the opposite side of his own dinghy was the one the failed assassins had left at the neighboring dock. Inside it lay two lumps in separate canvas bags, each properly weighted.

Rawls brought his dinghy alongside and took the other in tow. Some distance out on Penobscot Bay, he checked the depth sounder, found himself in nearly a hundred feet of water. He rolled each bag overboard and waited for it to sink, then he put a few holes in the other dinghy and watched it go under.

Satisfied with his evening's work, he returned to his dock, secured his boat, and walked to the house. He checked all his security checkpoints, then let himself in. Shortly, he searched his satellite directory for programs he had recorded, then settled in to watch a new season's debut of *Succession.*

Who were these people? he wondered of the cast. He wouldn't have any of them in his house!

CHAPTER 51

Stone received a call from Ed first thing the next morning.

"I thought you'd like to know, everything went as planned," Ed said.

"Glad to hear it."

Ed then filled Stone in on the details.

After breakfast, Stone and Dino took the Hinckley picnic boat out into the bay for a little fishing, though neither of them was a fisherman and both despised fish, except steaming on a plate.

Dino took a small device from his jacket pocket and fished for bugs of the electric kind on the boat, finding none. "I take it our friend operated successfully last night," he said.

"Twice," Stone replied.

"Is two all the opposition?"

"Let's hope so. I feel a little safer already, don't you?"

"If you say so," Dino replied. "You know, being a sworn police officer and all, I feel an odd satisfaction for the way this was handled. I've had a suspect or two now and then that I wanted to just pop in the head and put them out with the trash. Didn't you?"

"Yes," Stone said. "I always felt something between satisfaction and guilt when it was that way. It helps a lot that someone else is doing the dirty work."

"I think Ed enjoys the work," Dino said.

"It would be tough to train all your life for something, then have your skills wane over the years. This way, he gets to stay sharp."

"Do you think he has anything like a conscience?"

"I think," Stone said, "that his conscience gets at him when he sees someone he knows to be a bad, even evil, guy, walking around loose. That seems wrong to him, and he has a cure for it, called marksmanship."

"As long as he doesn't employ the cure in my jurisdiction," Dino said.

"You didn't seem offended when someone we know assisted an even more evil person into that awful cemetery in Queens."

"True. Have you had enough of non-fishing, yet?"

"Just about. I'd enjoy the boat ride more if it were a sunny day."

"Oh, that's right, you promised me some of those."

"I did, and you'll get them."

"I see Viv helped you into bed with Carly."

"She did?"

"Who do you think suggested to Carly to go outside and check on you? I hope you appreciate the effort."

"I do. I had just about put the idea out of my mind, until Carly brought it up."

"Bringing things up doesn't seem to be a problem for her."

"Definitely not. She sees things differently than most. It's the way she's wired. For instance, I know she really enjoyed watching Billy take out Alexei Gromyko."

"I hope getting laid was fun for her, too."

"I've been assured it was."

CHAPTER 52

Ed Rawls unloaded his groceries from his boat and carted them to the house. Once everything had been put away, he checked the fridge, and it seemed like he had more beef than he could eat in a week. He was about to call Stone and invite his crew to dinner when he stepped out onto the back porch to put away the cart and got a whiff of something on the breeze. Tobacco smoke.

Ed went back into the house and called Stone.

"Hey, Ed. How's it going?"

"I've got a surfeit of porterhouse in the fridge. How about you folks come down here this evening and help me reduce my stock?"

"Count us in," Stone replied. "Anything going on down there?"

"Funny you should mention that. I just got a whiff of cigarette smoke on the breeze."

"Uh-oh."

"Yeah, you'd better be carrying this evening. You bring the handguns, and I'll supply the long guns."

"Sounds good. What time?"

"Get here at six and watch the night fall."

"You're on."

Ed hung up the phone and went around the house, closing shutters and bolts and getting ready to receive fire after dark. Most assassins preferred that to daylight.

He leaned a rifle next to every window, and by that time alarm bells were going off. That would be Stone and party letting themselves in through the gates, using Ed's codes.

By six-fifteen, everybody was settled in Ed's living room with a drink.

"Any news from the opposition?" Stone asked.

Ed looked at his watch, then out the window. "I figure these people for cowards, so I'd say, half an hour after sunset, but I reckon we should hold the liquor until they've made their move. Let's douse all the lights, except that reading lamp, and get them in closer. I've laid out long guns and magazines for everyone. They'll be here soon enough, I reckon."

They doused most of the lights and waited, not talking.

The monitor on Ed's computer came alive, and he made

some adjustments. "Out on the dock," Ed said. "I figured they'd be too lazy to come ashore and walk through the weeds." Other screens on the monitor popped up.

"I see a boat out there, maybe a hundred yards," Dino said.

"And two men getting out of a dinghy at the end of the long dock."

"How you want to handle this?" Stone asked.

"With one round apiece, if I'm lucky," Ed replied. "Turn off that reading light, and I'll go out the back way and around the house."

"We'll wait until the firing stops before we crack a door," Stone said.

"I'll call with the all clear."

The lights off, Ed let himself out the back door and walked around the house, careful to avoid the crunch of feet on gravel. He stopped at the corner of the house, switched on his night sight, and swept the dock area. He found the two men still at the end, messing with ropes.

He hoped they were no better at shooting than with tying off dock lines. He took aim and waited for them to stand up. They finally secured the boat, then messed with their weapons.

"Light machine guns," Ed muttered to himself. "They're no better at being assassins than at boat handling."

The two men started creeping down the dock toward the house. Ed stood one of them up with a round and fired a second for insurance. The other fell flat on the treads, but Ed could still find him with the night scope.

He fired a single round into the top of the man's head, and watched him lurch, then lie still.

He got out his cell phone and called Stone.

"You still alive?" Stone asked.

"Yeah, but nobody else is. You light the grill, and I'm coming in."

"Consider it done. Like the song says, Knock three times."

CHAPTER 53

Ed Rawls put on an apron and started grilling three three-inch porterhouse steaks, and the evening became convivial. Dino manned the bar.

In the middle of somebody's long story, Carly raised a hand. Nobody paid attention to her.

"Hello?" she said plaintively. "Does anybody here smoke?"

Ed stopped talking and sniffed the air. "Everybody duck!" he yelled and led the way.

Glass began to break, and everybody ducked.

"There's more of them out there," Ed said. "My fault. I intended to leave everybody dead."

"Uzis," Dino said.

"Thank God for that," Ed replied, his cheek pressed to the floor. "Nobody move a hair! Complete silence!"

Carly crept across the floor on her belly toward the rear door.

"Goddammit!" Ed whispered. "I told you to be still!"

"I'll be still when there's a rifle in my hands," Carly whispered back and kept her course. She snagged a rifle and a magazine and rammed it home. "All set."

There was a small sound from the front deck, and a row of Uzi holes appeared in the door. Carly put four rounds through that door, then there was the sound of a falling body. "One down," she said. "Anybody joining me?"

"Hold it!" Ed said, then crawled to a front window and looked through a lower corner pane. "Carly, pass out weapons."

"Pass 'em out yourself," she replied. "I'm still working on your first order."

Ed crawled over to the front door, cursing under his breath. He started throwing rifles around the room, followed by magazines.

"Front door," Stone said, then put a burst in that direction.

"Two down," Carly said.

"It's four down, if you count the first two."

There was a shout from the direction of the dock, and feet could be heard pounding it. A moment later an outboard motor revved, and its sound began to fade.

Ed stopped and listened. "Gone," he said, finally.

"If you're that sure," Dino said, "stand up and look around."

Ed did so. "Everybody check the nearest window, but carefully," he ordered.

Somewhat reluctantly, the others stood.

"I'll be right back," he said, then walked out the rear door.

"I'm not moving until he doesn't get shot," Carly said.

"I'm right behind you," Dino said.

Ed Rawls burst through the back door, causing everybody to duck again.

"All clear," he said, "and good news, they took the bodies with them, so there's no cleaning up, except the glass. Plus, I remembered to pull the hood down on the steaks, so they should be glass-free."

CHAPTER 54

Eight men gathered around the dining table in a Rockland hotel suite, a few miles from Islesboro—Dmitri Asimov's inner circle.

Asimov rose from his place at the head of the table and rapped on the mahogany for attention. "I have news," he said, "of a sort."

"What does that mean?" someone asked.

"The second team returned two men short, and mission incomplete."

"They were our best men."

"No," said Asimov, "the two in our first team were our best men."

"How did this happen?"

"We sent the first two to take out Peter," he said, "but the marksman, Rawls, stopped them."

"So, we take out Rawls and go again with Greco."

"The second team was sent to take out Rawls," Asimov said. "It turns out that Rawls had more firepower at his disposal than we were prepared for. Our problem is that this is more of a military problem than a simple assassination. We need someone with a military background."

"The Sarge," someone said.

"Yes!"

"I'll get in touch and get him up here."

The phone rang, and the Sarge, who had actually been a captain before he was kicked out of the Marine Corps for trying and failing to murder his commanding officer, picked up the phone. "This is the sergeant."

"You know who this is?" a velvety voice said.

"I believe I do."

"I need you, and some more men, too."

"How many men and where?"

"There's a jet waiting for you at Teterboro Airport."

"How many does it seat?"

"Eight, but I don't think you'll need that many."

"How many targets?"

"Three, possibly more."

"Sounds like you need a little recce before the heads start to roll. Where are you?"

"Rockland, Maine."

"Is that where the targets are?"

"No, they're on a nearby island, Islesboro."

"So, we're talking about at least one boat. How about available weapons?"

"Bring what you need. Ammo, too."

"What else? An armored personnel carrier? A bazooka?"

"Equipment for night work."

"When do you need this done?"

"Last Thursday."

"Okay, let's hold it right there. This has all the makings of a first-class fuckup. I'll come up there and see what's involved, then I'll round up what I need."

"This needs doing right away."

"Then you're talking to the wrong man. I don't do right away. I just do it right. You want a referral? I know half a dozen people who can serve up a cock-up on demand, but then there'll be bodies everywhere."

"Oh, all right. Be at Teterboro, Atlantic Aviation, at nine tomorrow morning. We'll go over everything after you get here, then you can order what you need."

"See you later." Sarge hung up.

CHAPTER 55

Asimov squinted into the morning sky and picked up a black dot, which swiftly became an aircraft on final approach. It set down and taxied to the ramp. The airstairs door opened and a large man descended, followed by a smaller man.

Asimov shook the larger man's hand and said, "Hello, Sarge."

"Hello, Dimitri. My friend here is the Corporal."

Asimov shook hands with the smaller man, then the three of them got into an elderly Lincoln town car.

"Okay," Sarge said, "let's hear it."

"First, we get the ferry; we've got twelve minutes." The Lincoln shot forward.

The gates were just starting to close when they drove aboard.

"Is this the only way out here?" Sarge asked.

"For the public, yes. You may have whatever transport you need when the time comes. Right now, this is the best way to look things over, without attracting attention. There's a ferry back in an hour and a half we can catch. That should be enough time."

They drove through the village. "Stay out of the stores," Asimov said. "The island grapevine starts there, and you don't want to be on that radar."

"Gotcha," Sarge replied.

Asimov handed him a large-scale map of the island, then they drove on, until they stopped at a point where a driveway was interrupted by a large log.

"What the fuck?" Sarge said.

"This is where Rawls lives."

"Rawls?"

"He's their sharpshooter."

"Dimitri, let's move our asses out of here. I can already see four cameras. Drive a little farther and stop where you can."

Asimov followed his instructions.

"Now, Dimitri, what is Rawls's first name."

"Ed."

Sarge grimaced. "If you'd told me that on your first call, you'd have saved me a trip up here and yourself a lot of money."

"Why? What do you know about Rawls?"

"That he's the best shot and the smartest asshole the CIA ever produced. How many of yours has he killed already?"

Asimov looked uncomfortable. "Four."

"Okay, let's go back to the ferry."

"Don't you want to see the rest of the island?"

"No, whatever we do—if we do anything—starts and ends at Rawls's house. I don't want to be seen by anybody else on this island. Just park the car and go over the map with me."

Asimov did so, and Sarge made small marks on the map.

"This guy, Barrington, sounds familiar," Sarge said.

"He's an uptown lawyer, and his best friend is the police commissioner, Dino Bacchetti."

The ferry docked, and they got aboard for the return trip.

"Stay in the car, and keep the windows up," Sarge said. "Then let's go back to your hotel and talk there over a drink.

"All right," Sarge said when they had settled into the conference room. "Let's talk money. I initially saw this as a half-million-dollar job, but now we've got a powerful New York police official to deal with, not to mention a sharpshooter with the eye of an eagle and the brains of a genius. We're talking about two million dollars, in my Cayman Islands bank account tomorrow."

"You bury your dead and get them in and out of the Rockland airport," Asimov countered. "You eat all your expenses."

"Two and a quarter million," Sarge said.

"And you stay on until Rawls, Greco, and Barrington are dead."

Sarge considered it for a moment, then nodded. "Agreed."

"One last thing," Asimov said.

"What?"

"If it is possible, I would like to look Barrington in the eyes before you kill him. He has been an embarrassment to the family for too long. I want to see his fear, and make sure he knows it was me who ordered his death."

"That ups the risk to me and my people. Two and a half million and we have a deal."

The two men shook, then Asimov said, "You want dinner? We've got room service."

"I've got some calls to make," Sarge said, "but yeah, I could eat. And tell them to bring up a legal pad. I need to make a list."

CHAPTER 56

Ed Rawls rang Stone's doorbell and waited, fidgeting, on the doorstep until it was answered.

Stone ushered him in. "Lunch is soon," he said.

"I'll stay, but I'd like a drink, now, please."

Stone poured him a stiff Laphroaig. He had never seen Rawls so fidgety. "What's going on, Ed?"

"We've got new opposition," Ed said.

"What opposition?"

"I'm getting ahead of myself. I got a couple of calls this morning from people I take seriously, to say that questions are being asked about me."

"Okay, what else?"

"You don't understand, Stone. They know who I am."

"Why does that matter?"

"Before, I was a ghost, an old man on a pension. Now they know who and what they're dealing with."

"I see," Stone said. "And that will make them more cautious?"

"There's more. They've hired a specialist called the Sergeant."

"An assassin?"

"A whatever-it-takes guy, and he knows who and what I am."

Stone nodded, but he really didn't see what difference that made.

"It means they'll come after me, first," Ed said. "They know what I can do, and they won't want me in the way. They checked out my front gate earlier today."

"So, what does this mean to how we're operating?"

"It means I need to stop operating and start working on staying alive. I've got to get out of here."

Stone was speechless.

"I know that comes as a blow," Ed said, "but I'm no good to you dead."

Ed had a point. "Have you talked to Peter about this?"

"I sent him an email."

"When and where are you going?"

"As soon as possible, and I thought about my place in Virginia, then I thought about my girlfriend's house in England. Sarah's there now, and that's her preference."

"I'll call Mike Freeman, and see if he can offer you a ride over."

Stone went into his little office and called Freeman and made his request.

"Wheels up at eight AM tomorrow," Mike said. "Tell Rawls no weapons."

"Okay, Mike."

"You going, too?"

"No, I'm going to deal with it at this end."

"As you wish." They hung up.

Stone went back to the living room. "You're aboard a Strategic Services jet tomorrow morning, wheels up at eight. Mike says don't bring weapons."

"Got it," Ed said.

"Are you going to stay at your house tonight?" Stone asked.

"I haven't figured that out."

"I think we'd all better fly back to the city after lunch. I'll put you up for the night, and Fred can drive you to Teterboro in the morning."

"Thank you, I'm grateful."

"Go back and get your stuff after we eat."

"It's already in the car," Ed said.

Joan welcomed them all back to the city, and said, "We've had some lurkers."

"When and how many?" Stone asked.

"Two. They were around yesterday, then back this morning. They left about the time you landed."

"Well, either they've anticipated my flight schedule, or maybe they just got tired of watching an empty house, preferably the latter."

"What can I do?" Joan asked.

"Just keep the house looking uninhabited."

"Okay, no lights on the street side."

"And make sure we're always locked down," Stone said.

"You want me to shoot a couple of them?" Rawls asked.

"No, let's just let them wonder. If you need to leave the house, go through the garage tunnel or the garden gate. If anyone sees anybody who looks out of place, I want to hear about it immediately."

After Ed and Joan left the room, Carly said, "I'm not accustomed to being penned in."

"We'll try and keep you entertained," Stone said.

"Can you do that in bed? Right now, if you like."

"After dinner," Stone said. "I've got some catching up to do before that."

"I'll try to contain myself. In the meantime, I'll unpack and have a nap, so as to be fresh." She pinched him on the ass and headed upstairs.

They had an early-evening drink in Stone's study, then sat down to dine at seven.

Joan joined them. "The watchers have not returned," she said.

"I spoke to Greco a while ago," Stone said to Ed. "He said some people had a look at your property in Islesboro, then went away in a boat."

"What, they didn't torch it?"

"They knew that would make you mad," Stone said, "and they didn't want that."

"What did you use to do, Ed?" Carly asked.

"Pretty much what I do now, but in a more structured environment."

"And where was that environment?"

"Pretty much all over Europe and Scandinavia," Ed said. "Then I did a little stretch at the Atlanta federal prison."

"What for?"

"For doing my work a little too well."

"Tell her the truth, Ed," Stone said.

"Okay, the Russkies set me up in a badger game, then blackmailed me into pretending to give them classified information. I had a hard time selling that to my betters, so they yanked me off the street and sent me inside."

"Better let it go there, Carly," Stone said. "You don't want to know more."

"I always want to know everything," she said.

"That's a character flaw you're going to have to adjust."

"Funny, I never thought of curiosity as a character flaw."

"It is, in some circumstances."

"Okay, I'll drop it, for the time being."

Later, Stone had time for a lecture. "Let me explain something to you," he said to Carly.

"Good. I like explanations; then I can tell who's lying."

"Ed Rawls has spent his life in secrecy. He was a spy, after all, and an important one."

"Has Ed done something he's ashamed of?"

"We all have," Stone said.

"Not I."

"There will come a time, believe me, and you won't enjoy reliving that time in your life. Ed has spent a good portion of his life killing people, and he is proud that he did it well, but ashamed that he did it at all. When you grill him, it causes him pain."

"Oh, I get it."

"Try and restrain yourself. If you think you might cause him pain, ask me first."

"I can do that, I guess."

"I hope to God you can."

CHAPTER 57

Stone went down to breakfast the following morning because he wanted to be out of the bedroom before Carly woke and made further demands. Ed Rawls was already at the table.

Stone sat down. "Tell me about this Sergeant guy, Ed. He seems to cause you to become skittish. Why is that?"

"Sarge is an old Marine who got drummed out of the Corps for being too much himself."

"And what was he that frightened people when he was himself?"

"He was a homicidal maniac who enjoyed killing people. In fact, he almost killed his commanding officer, a man a lot of people would have enjoyed killing, but Sarge didn't mind telling people about how he had tried to eviscerate the man."

"I suppose that would cause a certain hesitancy in those who surrounded him."

"You could say that."

"Have you had dealings with him?"

"Not yet," Rawls replied, "and I'd like it to stay that way."

"Just how good an assassin is he?"

"As good as I am, I hear," Rawls replied. "But whereas I take satisfaction in doing my work well, Sarge apparently invests his job with a certain glee that is absent from mine. And frankly, that scares the living shit out of me. Which is why I try to be where he ain't."

"That seems to be working."

"Yes, but I can't tell for how long. I wasn't even sure that he was aware of my existence until I began hearing from friends who were hearing from their friends. Apparently, I'm no longer a casual target, but an assignment."

"And who is the assignee?"

"Those Russians who are supposed to be at Peter's beck and call, but who really want him not to exist. Not to mention me, who acted as his guardian."

"And me?"

"I don't know if he is even interested in you, but if I were you, I would behave as if he is. The Russians are not particularly fond of you."

"Greco is."

"Every rule has an exception. But it is a small one, in this case, and I wouldn't count on it saving you."

"How long do you think it would take Sarge to lose interest in us?"

"I wouldn't count on that, either."

"Then how do I stop him?"

"Kill him before he kills you."

"How?"

"Take your pick of weapons, and never be without one."

"Have you thought of going after him, Ed?"

"I've dreamed about it," Rawls replied. "I'll let you know when I come up with a plan."

"Please do."

Rawls sighed. "The plan may have to include using you for bait."

"If it does," Stone said, "I'd appreciate it if you try not to miss."

Stone saw Rawls off, then returned to the breakfast table for coffee. Carly was waiting for him.

"I had hoped for another round before breakfast," she said.

"I wanted to see Ed Rawls off and seek some advice from him."

"What advice?"

"How to thwart this guy, Sarge, and keep him from killing me."

"How?"

"By killing him first."

"How can I help?"

"Go armed and shoot first."

"You mean, seriously?"

"You should take him seriously, though I have no reason to think he knows you exist."

"So I should run from somebody who isn't chasing me?"

"Sort of like that. I do think you should stay away from me, until this thing is resolved."

"You *want* me to stay away from you?"

"No, no, that's certainly not what I want. It's just that I don't want to contribute to making your existence worse."

"You think I have a bad existence?"

"No, but I think it would make your life worse if someone is trying to kill you."

Carly thought about that. "I was in a similar position once, but I figured out how to resolve it."

"Tell me about it."

"Well, when I was about fifteen, there was this little shit named Bobby Haney in my class who thought he would amuse himself by making my life hell. He hit me a couple of times, and it hurt, but I was faster than he was, so I ran. Then I got tired of running."

"So, what did you do?"

"I got some books from the library on self-defense, read them, and practiced for about a month. I saw a couple of films on the subject, too. I got pretty good at it, I figured, but I still had a problem: my weight deficiency. I knew that if Bobby Haney hit me, he could hurt me so badly that I couldn't defend myself, so I worked on not getting hit."

"Did that work?"

"I had to find out using the real thing, not a dummy, so I sort of put myself in harm's way. I went to the gym after school, when I knew the wrestling team would be working out, and I waited until they had finished and were leaving the mats, then I walked over to where Bobby Haney was toweling himself off and just stood there. He was the last to leave, but then he saw me. He said something like, 'You little cunt.' So I spat in his eye. I'm still a pretty good spitter."

"And what did Haney do?"

"What I knew he would do. He ran at me. I did a little sidestep and tripped him. He fell off the mat, onto the hardwood floor and landed on his chin. I figured that was as good as hitting him there, because he bit his tongue."

"Okay, tell me why you're still alive."

"When Haney got up, I kicked him in the ribs, and he went down again. He kept doing that, and so did I. Pretty soon, he couldn't get up anymore; he was sort of helpless. So, I walked slowly over to where he sat, pulled his chin up, and hit him in the nose, as hard as I could. I felt the cartilage break. He spouted blood, but he still couldn't get to his feet. I could hear feet behind me, running toward me, so I turned and dodged a couple of wrestlers, tripped one of them, who fell into a pool of Haney's blood. The other one stopped, and just stood there, no doubt wondering whose blood he was seeing."

"What happened then?"

"I just looked at Haney and said to him, 'I don't want to see you again, except in class. Anywhere else, and I'll hurt you.'"

"And that was it?"

"I never saw him again, except in class."

"That's a good story, Carly, but we're not dealing with school bullies here. So please don't try that, if anybody bothers us."

She shrugged. "As you wish."

CHAPTER 58

Stone got a call from Bill Eggers, his managing partner. "I need you in my office for lunch with a prospective client," he said.

"Sure, who's the client?"

"Name of Peter Greco. He's the CEO of a conglomerate, made up of half a dozen companies. He reckons they'll spend about three million a year on legal fees."

"See you at lunch." He didn't mention that he knew Greco and that he had declined to represent his group, something that Greco had said he understood and would respect.

Stone was on time and accepted a glass of tonic water, as he sat at the table.

"Greco is late," Eggers said.

"How did he come to you?" Stone asked.

"He knew a guy I was at Harvard with."

"Not much of a recommendation."

"He mentioned three million in fees. That was enough of a recommendation."

"Right," Stone said.

An hour later, Eggers left the conference room then came back. "I guess he isn't going to show. His secretary said he had left his office in plenty of time to make our lunch." Eggers absently speared a slice of quiche and started eating it. "When are you going to send Carly back to us?"

"Maybe never. She's turned out to be very useful. I don't have to use a dictionary anymore."

"She has a big vocabulary?"

"Well, yes. When she was twelve, she says, she memorized the *Oxford English Dictionary*."

"How much of it?"

"All twenty volumes of it."

Eggers stared at him. "You're kidding."

"I kid you not. Just ask and she'll give you the spelling, definition, and etymology of any word in it, though I rarely need the etymology."

"I don't think you should let any of our clients know that," Eggers said.

"Why not?"

"She'll scare them to death."

"I hadn't thought of that."

They finished eating, and Stone stood. "Thanks for the lunch. See you around. Let me know if you want to reschedule."

Stone left the building and walked slowly back to his house, crossing the street now and then and watching the reflections in shop windows to be sure he wasn't being followed. He seemed in the clear.

Back at his office he greeted Joan, who was hard at work, typing something on her computer. He was still hungry, so he walked back to the kitchen and found a pickle, which had just the right flavor. When he reached his office, the door was open, and a man was seated opposite his desk, his back to the door.

Stone returned to Joan's office. "Who's the guy in my office?"

"What guy?" she asked, mystified.

"The one in my office. Who is he?"

She rose and walked ahead of him to his office, strode in and said, "I'm sorry to keep you waiting, sir. Mr. Barrington has just returned from a meeting."

Stone followed her into the room. "Good afternoon," he said.

Joan was standing next to the man's chair, looking down at him. "Sir?" she was saying, then she froze.

Stone stopped in his tracks. "What?"

"I think he's dead," she said.

Stone walked to her and looked down at the man. He was Peter Greco, and Joan was right.

"Well," he said, "at least he has a good excuse for standing Eggers and me up for lunch."

Carly walked into the room. "Hi," she said. "Something smells dead."

"Funny you should mention that," Joan said. "I guess our air-conditioning isn't working well enough."

"Joan," Stone said, "call 911, then get me Dino."

"Why don't we skip a step and just call Dino?" Joan asked.

"Dino doesn't like skipping that step," Stone replied. "He likes to be the second to hear."

Joan made the calls, and yelled, "Dino on one!"

Stone picked up. "I just came back from meeting with Eggers at Woodman & Weld and found a stiff waiting for me in my home office."

"Which one? I know a lot of stiffs."

"This one is both stiff and cold and is starting to impair our atmosphere," Stone said.

"Anybody we know?"

"Peter Greco."

"Why am I not surprised? Did you call 911?"

"Of course."

"When the detectives get there, tell them I'm on the way and not to touch anything. Don't you or yours touch anything, either." He hung up.

Carly was standing next to the desk, staring at Greco's corpse. "Why would they kill him somewhere else and leave him here?"

"I don't know. Maybe their dumpster was full."

"Right, that makes a lot of sense."

"Or maybe as a warning."

"What kind of warning?"

"The 'this could happen to you' kind of warning."

"Oh. You think it could be the Sarge guy?"

"It would be foolish to think otherwise."

"He sounds like an unpleasant person."

"He does, at that."

"I wonder why he thought he had to prove it?" Carly mused.

There was a low moan from a police siren outside, and Joan let in two detectives—as usual, one Stone knew named Kelly, and a younger one.

"Hello, Stone," Kelly said. "I hear you've got some business for us." He gestured toward the other guy. "This is my partner, Smith."

"What kind of name is that for a cop?" Stone asked.

"Call me Smitty," the younger man said.

"Smitty, it is," Stone said. "Call me Stone, if this is going to be a pleasant meeting."

"Just a few questions," Smitty said, taking a seat near the corpse.

"Shoot."

"Did you kill him?"

"Nope. He was like that when I got home from a meeting at my law firm."

"How long has he been dead?"

"Beats me."

"About two hours," Carly said.

Stone introduced her to the two cops.

"How do you figure?" Smitty asked her.

"I have a nose."

"A better one than mine," Stone said.

"Okay, let's start with where everybody was two hours

ago. We gotta do something until the ME gets here and tells us if we got a psychic on our hands." He nodded at Carly.

"My advice is to take her word for it," Stone said. "She's not often wrong."

"Hardly ever," Carly added. "I worked in the New Haven morgue nights and weekends as an undergrad at Yale. At the morgue, they called me 'the Nose.'"

"Okay," Smitty said.

A man wearing scrubs and carrying a satchel entered the room. "I'm Dr. Carson," he said, looking at Greco and sniffing the air. "About two hours, I'd say."

"Can you get him out of here before *I* start smelling him?" Stone asked.

Dino entered the room. "Okay, how long?" he said to the ME.

"About two hours."

"Carly had already made that estimate," Stone said. "If it wouldn't be too much trouble, I'd appreciate it if he were moved before he stinks up the whole place."

Dino looked at the ME and jerked a thumb.

"Whatever you say, Commissioner," the man replied, and made the same gesture to the two men pushing the gurney. The atmosphere began to improve.

The ME reentered the room, ten minutes later. "Preliminary cause of death, a small puncture wound in the neck, near the base of the skull," he said.

"Ice pick," Dino said.

"Possibly. Doesn't appear to be a small-caliber gunshot: no powder burns."

"Any other news?" Dino asked.

"Not before I've got him splayed on a table," the ME said. "Can you contain yourselves for a few hours?"

"We can," Dino said. "There's not going to be much news, anyway."

"You're a very nice policeman," the ME said. "Good day." He closed his bag and left.

"I'll bet nobody's ever called you a 'very nice policeman' before," Stone said.

"Not outside of the sack," Dino said. "My wife sometimes calls me that after a bout in the hay."

"What other affectionate terms does she use?" Carly asked.

"Ask me when I've had a couple of drinks," Dino said to her.

"I wish Rawls were here," Stone said, looking at his watch. "I can never remember what time it is in England."

Dino glanced at his watch. "Middle of dinner, I should think."

Stone's phone rang. "Yes?"

"It's Rawls. We had a good flight and Sarah has a pair of shotguns that we've cleaned, oiled, and loaded with buckshot."

"I'll bet Carly can smell the gun oil from here," Stone said.

"Oh, stop it," Carly said.

Rawls laughed. "Has she been confounding you all with her powers?"

"She has. She just nailed the time of death of a corpse in my office at home, ahead of the ME."

"Good going, Carly!" he hooted. "Who's dead?"

"Peter Greco, ice-picked in the neck and delivered fairly fresh."

"Oh, shit. I got out just in time, didn't I?"

"I hope so. And I hope you're not next, because I'm in line right after you."

"You're just going to have to kill the next Russian in line, since I'm not there to do it for you."

"I would, but I don't think our police commissioner will let me."

"Maybe somebody in his shop can tell us who's in line after Greco. He'll be the guy who hired the Sarge. I don't have a guess, I'm afraid."

"I'll ask."

"Do you want me to come back tomorrow and watch your ass?"

"Don't bother. Strategic Services has that view of me."

"I'm always available, once I've relaxed. I got relaxed on the flight over, when nobody shot us down."

"I'll keep you posted."

"You can only do that if you're alive. If you fail to call, I'll assume the worst. Find the Sarge and kill him."

"It would save us all a lot of time if you'd just do that anyway."

"Maybe you can talk Dino into doing just that. He

has more resources at hand than I. Tell him somebody took a shot at Viv; that'll set him off."

"Yeah, but if he finds out I'm lying, that'll set him off, too, and at me."

"I leave it in your capable hands, Stone, for the moment. Bye." Rawls hung up.

To Dino, Stone said, "Do you or your people have a guess at who is next in line to take over now that Greco is dead?"

"If you asked me a week ago," Dino said, "I would have said yes. But then Greco got the job. Nobody saw that coming. So, who knows now?"

"Take Greco out of the equation. Who should have been given the job instead of him?"

"Let me make a call." Dino pulled out his phone and stepped out of the room. When he returned, he said, "One of three men."

"And they are?"

"Igor Krupin, Dmitri Asimov, or Gregor Dryga. Krupin's at the older end, put in his time, knows where all the bodies are buried. Literally."

"I've heard of him. He's been around in the background for a while."

"Asimov is one of the younger upstarts. Hot-tempered, doesn't like to wait around for things to happen, the kind of guy who likes to fix problems with a sledgehammer. It doesn't matter how small they are."

"He sounds lovely," Stone said.

"I'll give you his number. You can have drinks together," Dino said. "The last guy, Dryga, is known as the

Bean Counter. He's the guy who took over the CFO duties from Greco. Before that, he handled logistics for several of the family's businesses. He's smart, calculating, and obsessed with details."

"As supervillain names go, the Bean Counter is not great."

"Maybe he needs a better PR guy."

"What's your gut on who has the inside track?"

"Given how quickly they brought in the Sarge, I'd lean toward Asimov."

"Do we know where he is?"

"Why? You really interested in grabbing that drink with him?"

"Just the opposite. If I know where he is, I know where not to be."

"I thought you'd be more concerned about the Sarge."

"If you know his whereabouts, I'll gladly take that information, too."

CHAPTER 59

Ed Rawls found his girlfriend Dame Sarah Deerfield waiting for him in living room.

"Ready?" she said and stood.

She had attired herself in a red dress that hugged her in all the right places. Ed took a moment to admire the view before saying, "It might be better if we eat in tonight."

Her eyebrow crooked. "Problems in New York?"

"Greco is dead." He didn't need to tell her the rest as he'd informed her about what had been going on when he'd first arrived.

"Which means you're next," she said.

"Me or Stone. Or, depending on how ambitious they are, both of us at the same time."

"I don't like the sound of any of that."

"It doesn't exactly fill me with joy, either."

"Shall I call my friends at Scotland Yard? I could have two dozen officers surrounding my house within the half hour." Sarah was the retired chief of the Metropolitan Police in London.

Ed considered the suggestion, then nodded. "Make the call but let me check the area before you have them show up."

She pulled out her phone, and he headed upstairs.

Ed checked Sarah's security cameras on the monitor in the bedroom and saw no obvious signs of trouble. But if the Sarge was as good as his reputation, there wouldn't be.

Ed opened his suitcase. While he had followed Mike Freeman's instructions to the tee and had brought no weapons with him on the flight from Teterboro, he *had* packed a few items typical travelers would not have. Tucked among the stuff in his bag were a pair of binoculars, with both available light and night vision modes.

He grabbed them, made his way onto the roof, and crawled to a position from where he could survey the street. He used available light mode first and scanned as far as he could in either direction. Nothing stood out. He then switched to night vision and paid special attention to the areas where little light reached.

His gaze stopped on a person crouched behind a car near the end of the block, who was leaning out just far enough for a sliver of their body to glow green in the binoculars.

"Who are you?" he whispered.

He watched the figure for a few moments, waiting for

the person to move. But whoever it was just peered down the street toward Sarah's house.

Ed crawled back until there was no chance he'd be seen, and called Sarah.

"The police are all set," she said, "just waiting for my go ahead."

He told her what he'd seen.

"I'll send a few officers in on foot first," she said.

"Good idea."

Ed moved back to his former position and trained the binoculars on the lurker. Soon, two of London's finest came around the corner on the same side of the street the lurker was hiding. A moment after that, two more appeared at the corner on the opposite side.

Hearing the steps, the lurker dropped to the ground and scooted under the car. This did not fool the officers. They surrounded the vehicle, then one of them shouted something Ed couldn't make out.

After a moment, an arm appeared from under the car, then the lurker's head and shoulders. As soon as the officers saw that the suspect—who appeared to be a man—was unarmed, they rushed in and yanked him to his feet.

Ed called Sarah. "Send in the rest of the troops."

When he rejoined her downstairs, a policeman was with her.

"Ed, this is Chief Superintendent Rogers," Sarah said. "Chief Superintendent, this is Ed Rawls. The Chief Superintendent is in charge of the officers who will be watching the house."

"Thank you for coming," Ed said. "What's happening with the person you apprehended down the street?"

"He is outside," Rogers said. "Dame Deerfield thought you might wish to speak to him."

"Very much."

"This way."

The lurker was in the back seat of a police car, eyes wide in fear. He was younger than Ed had expected, much younger—no more than eighteen or nineteen.

Rogers nodded at one of the waiting officers, who opened the door, and pulled the lurker out.

"Please, this is some kind of mistake," the kid said. "I haven't done anything wrong!"

"Then why did you hide under the car when the police arrived?" Ed asked.

The kid blinked, confused by Ed's American accent or the fact that he wasn't wearing a uniform or both. "I—I didn't know it was the police. I just heard them coming toward me."

"And your first instinct was to hide. Why?"

The kid looked away, clearly not wanting to answer.

"What's your name?" Ed asked.

"Christopher. Christopher Bedford."

"Christopher, the more you cooperate, the easier things will go for you."

Ed's words obviously had the opposite effect from what he intended as the blood drained from the kid's face. "I didn't do anything wrong."

"If that's the case, then just tell us what—"

"Christopher?" a female voice called from across the street.

A young woman about the same age as Christopher ran to the police car. "What happened?" She turned to Rogers. "Why are you holding him?"

"You know this lad?" Rogers asked.

"Yes. He's my . . . my friend."

Ed instantly understood what was going on here. But before he could say anything, a man stepped out of a house across the street, took in the scene, then called, "Caroline, what's going on? Why are the police here?"

The girl took a step away from the car. "It's nothing." She shot Ed and Rogers a quick look, pleading for them not to say anything, then she turned and jogged across the street.

To Christopher, Ed said, "I take it her parents don't know about the two of you."

Christopher shook his head.

"I'll let you handle this," Ed said to Rogers and then returned to the house.

He was glad it was a false alarm, but for some reason that made him feel more concerned rather than less.

"How about a drink?" Sarah said.

"How about more than one?"

A man on a rooftop, three blocks down from the house where Ed Rawls was staying, lowered his binoculars and called the preprogrammed number on his phone.

"I've scouted the location," he reported.

"Your assessment?" the Sarge asked.

"It's not worth the trouble."

"Explain."

"I don't know who this guy is, but he must have friends in high places. He's got at least twenty Metropolitan Police officers protecting this place. Maybe you could get to him, but I wouldn't bet on your chances of getting away after."

"You're sure?"

"It's my job to be sure. Feel free to hire someone else to check it out if you want, but the answer is not going to be any different."

"Dammit," the Sarge spat. "All right. I'll have the second half of your payment wired to your account."

"Pleasure doing business with you."

CHAPTER 60

That evening, in New York City, the family council gathered at the restaurant in Little Italy, in the room that had until recently been Alexei Gromyko's office.

Asimov arrived first, and purposely took the seat at the head of the table.

When Igor Krupin walked in several minutes later, he was not pleased. As the eldest, the position of chairing the meeting should have been his. "I believe you are in my seat."

"Your beliefs are not important to me," Asimov said.

Krupin's gaze hardened. "Move."

"I like it here, thank you." Asimov motioned at the other chairs. "There are plenty of other places for you to choose from." He turned away, ending the conversation.

The rest began filing in. Those who weren't part of Asimov's cabal looked askance at Asimov's place at the head of the table, but none challenged him. On the other hand, his friends smiled in approval as they filled the chairs nearest him. The only one who looked neither happy nor appalled was the Bean Counter. He was his usual calm, unreadable self, and took a chair midway down the table, right between Asimov's friends and foes.

"I am sure you have all heard the news," Asimov said, after everyone had settled, "but in case some haven't, Egon Pentkovsky—or, as he preferred, Peter Greco—is dead. I know none of us expected to be discussing leadership of the family so soon after we just did, but here we are."

"Do we know who did it?" one man asked.

"It has to be the same person who had Alexei killed," another said.

"The lawyer? But I heard Greco was found in Barrington's office," a third man said. "Barrington isn't stupid enough to have killed him there, is he?"

Everyone started talking over each other, asking questions, throwing out theories, and shouting suggestions for retaliation.

Asimov let it go on for a minute, and then slammed his hand on the table. "Enough."

The room quieted.

"I already have people looking into the circumstances of his death," Asimov said. "Have no doubt, when the perpetrator is discovered, he will be dealt with in no uncertain terms."

"Need I remind you how he is dealt with is not your

decision to make?" Krupin said. "It is that of the new head of the family."

Asimov smiled, menacingly. "So it is. Then we should decide who will make that decision."

"By rights, it should go to Igor," one of Krupin's allies said. "He is the senior among us."

"Is that the will of everyone?" Asimov asked.

A few heads nodded, but from the crease in Krupin's brow, their number was not as large as he had expected.

A few seats away from Asimov, one of his friends cleared his throat. "I think the job should be yours, Dmitri," he said, looking at Asimov. "You are a man of action, and that is what's needed now."

Several others voiced their agreement. More, Asimov noted, than had for Krupin.

"I'm flattered you think that," Asimov said.

"We should put it to a vote," the man said.

"Is there anyone else we should consider?" Asimov asked, wanting to sound egalitarian, when he was anything but.

He looked around the table. The only other name that might be thrown out would be that of the Bean Counter, but the man apparently had no plans to nominate himself, nor did anyone else speak up.

"All right," Asimov said. "It appears the choice is between Igor and myself." He turned his attention to the man who'd put Krupin forward. "Kazimir, would you be so kind as to conduct the vote?"

"Of course."

It came down to a single vote, that of the Bean Counter. Asimov had no idea which way he would go. The man had always been hard to read. When Krupin's and Asimov's names had been put forward, the Bean Counter had sat stoically and not displayed support for either man.

If Asimov didn't win, he and those who supported him had already agreed to take control of the family by force. He inched his hand toward the butt of the pistol hidden under his jacket.

Kazimir said to the Bean Counter, "Mr. Dryga?"

The Bean Counter stared at the table, his lips resting against his clasped hands. Finally, he leaned back, and said, "Dmitri Asimov."

Though Asimov had known that no matter what happened, he would be in charge at the end of the meeting, achieving his goal was still a shock.

"The council has decided," Kazimir said. "Dmitri Asimov, the leadership of the family is yours."

One by one, the members of the council pledged their support, even Krupin, though his face was ashen, like he might drop dead from a heart attack at any moment.

Asimov thought that was probably something he should consider facilitating. Krupin had served the family well, but going forward, the man would be a thorn in Asimov's side. Best to usher him into permanent retirement. Perhaps in one of the empty plots near where they'd just buried the Greek.

Asimov smiled at the idea.

The Bean Counter headed to his office in the back seat of his town car, Korolev driving. The lieutenant's gaze kept flicking to his boss in the rearview mirror, then back at the road.

"Go ahead and ask," the Bean Counter said.

"Is it decided?" Korolev asked.

"It is."

"And?"

"Asimov will lead the family. For now."

"It was as you thought."

"Yes."

Asimov may have believed the details of the coup he'd just orchestrated were only known to those close to him, but the Bean Counter knew the upstart was the one who had ordered Greco's death. In the Bean Counter's role as CFO, he had his fingers in every aspect of the organization, and his appreciation of underlings who did their jobs well had gained him a network of informants.

His vote for Asimov had not been made because Dmitri was the right man to lead the family. He was not. The Bean Counter had thrown in with him for two reasons: first, if he had chosen Krupin, the meeting would have ended in violence, and possibly his own death; and second, he had little doubt Asimov's vendetta against Stone Barrington and Ed Rawls would backfire. When it did, the organization would need a new, competent leader such as himself. And if things didn't go belly-up, his vote meant he was the one who'd put Asimov in power, and Asimov would be appropriately thankful.

"So, what is the plan now?" Korolev asked.

"For the family? Or for us?"

"Isn't everything we do *for* the family?"

"It is," the Bean Counter said with a smile. "The plan is we wait and see who's still standing at the end."

"Then you'll make your move."

"Then *we'll* make our move."

CHAPTER 61

Stone and Carly were on the way to Clarke's to meet Dino for dinner when Stone's secure cell phone rang with a call from Holly Barker.

"I had been thinking about paying you a visit this week," she said, "but the Secret Service tells me they'd rather I wasn't around you right now."

"That's a nice way of putting it," he said.

"Well, what they said was closer to 'you cannot go anywhere near Stone Barrington until further notice.' I take it you are under threat of death again."

"Unfortunately."

"Who is it this time? Wait. Don't tell me. If I really want to know, I'll ask Lance. Can I at least assume it's not a jealous husband?"

"You can."

"That's good to hear, but I'm disappointed you are unavailable to me, and for an indeterminant time. You are a wonderful stress reliever."

"I hope I'm good for more than that."

From the seat beside him, Carly gave him a curious look.

"You are, have no doubt about that. It's just a woman in my position has few chances to get comfortably naked with someone who knows how to please them."

"You could always try someone who doesn't know how to please you, and consider it a chance to train them."

"Who has the time for that? Anyway, I just wanted to tell you to try to stay alive. The mood of your president depends on it."

"I'll do what I can, Madam President."

He hung up.

"Did you just get a booty call from the president of the United States?"

"I plead the Fifth."

Carly slid her arm through his. "You should call her back and tell her you are currently occupied."

"No need. She's not coming. The Secret Service isn't keen about the Russian mob's current interest in me."

"Huh."

"What?"

"It's just . . . Is it wrong that I'm suddenly grateful someone wants you dead?"

"Carly, that mind of yours is a dangerous place."

"You are not the first person to say that, but I can never tell if it's a compliment or not."

"Pretend it is, and we'll all be happy."

Dino was at the bar when they arrived.

"I thought you were traveling with an entourage," Dino said.

"We are," Stone said. "There are at least six of Mike Freeman's people scattered throughout the restaurant, and four more at the front door, who arrived with the two cars that escorted us."

"Mike is not messing around."

"He never does."

They had one drink and then were shown to their table. During appetizers, Stone's secure phone rang again.

"I never knew the president was so persistent," Carly said.

Stone looked at the screen. "She can be, but this isn't her." From the lack of an identifying number, he knew it was Lance Cabot. Given that Stone had threatened his job during their last conversation, he wasn't sure what to expect. He accepted the call. "Good evening, Lance."

"I heard about the corpse in your office," Lance said. "Is business that bad?"

"For Peter Greco, it is."

"Indeed." Apparently, Lance was taking the pretend-it-didn't-happen route. "Speaking of Greco, your Russian friends have selected a new leader to replace him. A man named Dmitri Asimov."

"I don't think Dino even knows that yet."

"Of course, he doesn't. Is he aware the Russians are working with an assassin known as the Sarge?"

"That he does know. We're assuming he's responsible for Greco's demise."

"You would not be wrong."

"If you have proof of that, I'm sure Dino would appreciate it if you shared."

"Enjoy your dinner, and tell Carly I send my greetings." Lance hung up.

Stone told Dino and Carly what Lance had said, less the greeting for Carly.

"So, the family went with the hothead," Dino said. "That was my least favorite option."

"Perhaps if they had consulted with you first, they would have gone in a different direction," Stone said.

"I'll give them a call next time."

Stone woke the next morning feeling more tired than he had when he'd gone to sleep. In response to Holly Barker's call the previous evening, Carly had seemed determined to prove that there were more reasons for the president to stay away than just the Russians.

She had made her point, multiple times, much to Stone's pleasure at the time and chagrin now. Thankfully, he was alone in the bed, so there seemed no immediate threat of another demonstration.

He showered, dressed, and headed down for breakfast, expecting to find Carly there, but the dining room was empty.

He found Helene in the kitchen. "Have you seen Carly?"

"She wanted me to tell you she had a meeting at the office and would be back at lunchtime."

"She didn't leave alone, did she?"

"Fred took her. And before you ask, she also said to tell you she was carrying."

That was a relief.

"Would you like your breakfast now?"

"Yes, please."

Stone ate and then made his way to his office. He had it in his mind to call Ed Rawls and tell him he'd reconsidered, and would appreciate it if he came back to the States to watch Stone's back. He had just sat down when Joan walked in, carrying a box covered in gold-and-black wrapping paper and matching ribbon.

"What's this?" he asked.

"A present for you," she said.

"From whom?"

"Unknown. There was no card, and the man who delivered it didn't know, either. Would you like me to open it?"

"Please."

She set it down, untied the ribbon, and peeled back the wrapping. She used a pair of scissors to open the box underneath.

"Oh," she said, peering inside.

"Oh good, or oh bad?"

Joan reached into the box and pulled out a desk clock, housed in mahogany. "This is beautiful," she said. "I think it's handmade."

She set it on the desk and Stone looked it over. It was

exquisite. The kind of clock that would cost several thousand dollars. While clients often sent him tokens of gratitude, few were of this caliber, and he could think of no current client who had reason to gift him in this way.

Joan looked back into the box and pulled out an envelope. "Ah, I guess there *is* a card."

She handed it to Stone. He opened it and pulled out a brochure for a mortuary. On the front was taped a note that read: *Best to get your affairs in order sooner than later.* There was no signature.

"That doesn't look like a card," Joan said.

"Because it's not." He turned it so she could see.

"I wasn't aware we had a mortuary as a client."

"As far as I know, we don't."

"Then why would they send you this?"

"I doubt they had anything to do with it," Stone said. "This is from someone trying to send me a message."

"What message?"

"Something about time, and how little of it I have left."

"That's presumptuous."

"Get Dino on the line."

She started to leave.

"Joan?" he called before she reached the door.

When she looked back, he pointed at the clock and the box, then motioned for her to take them out and put them somewhere far from either of their offices. It was possible the clock was just a clock and the box just a box, but he would feel better after Mike Freeman's people checked them both for hidden devices.

Joan put the clock back in the box and carried it out.

A couple minutes later, she stuck her head into the office. "Dino on one."

"And the box?"

"In the garden shed."

Stone picked up the phone. "I just received an interesting package."

"Another stiff, like yesterday?"

"No stiff. Just an expensive clock and a brochure for the Dalby Family Mortuary."

"I hear they do good work."

"That makes me feel so much better," Stone said. "There was a note with the brochure." He read it to Dino.

"Sounds like someone thinks you won't be breathing for much longer."

"One guess who that would be."

"The Russians," Dino said, "by way of the Sarge, most likely. Am I close?"

"You got it in one."

"What I don't get is if they're planning to kill you, why send you messages like Greco's corpse and this fancy clock, instead of just doing it?"

"Aside from the fact that I'm glad they are taking their time, I think they were at first trying to scare me. Now, they may be attempting to get me to do something."

"Other than to avail yourself of the Dalby Family's services?"

"Other than that."

"Like what?"

"Ask for Ed Rawls's help."

"That makes sense. Easier to kill you if you're together."

Stone had been close to doing just that.

"Maybe I should issue an APB for the Sarge," Dino said, "as a person of interest in Peter Greco's murder."

"He'd never let himself get caught."

"No, but he'd have to be more cautious. No more freely moving around."

That was true. "Good idea."

"I have them on occasion."

"So I've heard."

Joan peeked her head into the room. "Ed Rawls on two."

"I'll let you two bond over your shared destiny," Dino said, and hung up.

Stone switched to the other line. "Hello, Ed. Everything all right?"

"So far so quiet today," Ed said. "The Metropolitan Police have the area around Sarah's house locked down. I have to say, though, I'm not used to being under police protection. It's a little like being back in prison, just with a nicer cell."

"And cellmate," Stone said.

"That also is a plus. But I am starting to wonder if you might need me back there."

"I'd advise against it."

"Has the threat level to you gone down?"

"Quite the opposite." Stone told him about the package, and his and Dino's theory about the Sarge baiting Ed to return.

"As much as I hate to say this, you're probably right. The only way they can get to me here is by shooting an anti-tank missile at the house, from several blocks away."

"Do you think that's a possibility?"

Ed said nothing for a moment. "A possibility? Yes, but not a large one. Obtaining and moving a missile around the U.K. would be more difficult than back home."

"You sound as if you feel safe there," Stone said. "I recommend you stay."

"The Sarge isn't going to wait long for me to come back before he decides to make a move on you anyway. What are you going to do?"

"I'm working on that," Stone said.

"My suggestion," Ed said, "work faster."

CHAPTER 62

That afternoon, someone knocked on Stone's office doorway. "Mr. Barrington, do you have a moment?"

Stone waved in the specialist Mike Freeman had sent to check out the desk clock.

"You were right to be concerned," she said. "There are two bugs in the clock and one embedded in the packing material in the box."

"I'd like to say I'm surprised."

"Maybe this will do the trick. Whoever sent the clock wasn't playing around. The bugs employ the latest tech and are not cheap. Their selling point is that they are undetectable. And to most scanners, they would be. Lucky for you, Strategic Services also uses the latest tech. But even then, I almost missed them."

"I appreciate your efforts."

"Would you like me to take them with me when I leave?"

Stone ignored the specialist's question and asked one of his own. "What kind of range do the bugs have?"

"If you're worried that they might pick up our conversation here from the shed outside, don't be," she said. "It's nowhere near that sensitive. I would avoid having sensitive conversations in your garden, however."

"That's what I was hoping you'd say. Thank you. Let's leave them in the shed for now. If I change my mind, I'll call Mike."

"Understood." She left.

Joan popped in a moment later. "So?"

"So, be careful what you say in the garden," he said. "And tell Carly, too."

"Tell me what?" Carly asked from the doorway.

"To watch what you say."

"You've already told me that, multiple times."

"This time I mean when around the shed in the garden."

"That's oddly specific."

Stone explained to her about the clock.

"Okay, that makes sense now," Carly said. "But I don't understand why you're keeping it around."

"Because it could provide us with an opportunity."

"What kind of opportunity?"

Stone tapped his temple. "I'm working on that."

"You should work faster."

Before he could respond, Joan said, "Oh, I almost forgot," and laid a white envelope on his desk, which had nothing written on the outside.

Stone leaned away from it, as if it were a snake. "Who is it from?"

Joan rolled her eyes. "Stop overreacting. If I didn't know the sender, I would have told you first. This came from the office." She handed an identical envelope to Carly. "And this one is for you."

Carly opened hers first. "A party. How fun."

Stone pulled the card out of its envelope. It was an invitation to a dinner being thrown by the New York State Bar Association. The fete was occurring the following Friday, and would be honoring a trio of attorneys for their charitable work.

Stone said to Joan, "Send the usual 'thank you for inviting me, but I will be unable to attend.'"

"Perhaps you should read the entire invitation first," she said.

"One of the celebrants is Bill Eggers," Carly said.

"Or Carly can tell you," Joan said.

Stone looked at the invitation again, and Bill's name was there in black and white. "I guess there's no way I can get out of this, is there?"

"Not unless you want Bill reminding you of the fact every time you see him."

"I do not."

"It can't possibly be that bad," Carly said.

"Sitting around listening to a bunch of lawyers talk about lawyering is not my idea of a good time."

"I know something that should help."

"What?"

"We can go as each other's date," she said.

"I accept."

The next day, another gift arrived. This time it was a basket containing a bottle of cabernet sauvignon and a bottle of sauvignon blanc from Screaming Eagle winery in California, worth an easy ten thousand dollars. Accompanying the bottles was a brochure from Riegel Mortuary in Queens.

As before, three bugs were found. Thankfully, none were hidden in the corks of the bottles. Stone would have hated to waste such fine wines. Instead, they were located within the weaved fibers of the basket.

Stone let the basket remain in his office for an hour while he made innocuous calls to clients, then asked Joan to put the wine in the cellar and discard the basket, a reasonable action that those listening in would be unlikely to question.

The day after that, a third gift was delivered.

The modicum of subtlety the sender had been using to this point was gone. This gift was entirely contained in an envelope. In addition to the expected brochure—this one from Grob & Grob Mortuary on Staten Island—there was a gift certificate for a top-of-the-line bronze casket. On the upside, there was no room in the envelope for another bug.

Stone shredded it and burned the remains in the fireplace.

That afternoon, as Stone continued to contemplate the Sarge problem, something Ed Rawls had said kept playing in his mind.

The plan may have to include using you for bait.

This, in turn, spawned an idea.

Stone didn't relish the thought of being a worm on a hook, but all the other schemes he had come up with had been even less likely to flush the Sarge out.

He called Teddy Fay and told him what he was thinking.

"That's not bad," Teddy said. "Not bad at all. It *is* a little short on details."

"Which is why I called you," Stone said. "Details are your specialty."

"Is Carly around? I found spitballing ideas with her last time to be very productive."

"I don't know if that's a good idea. I'm worried that she'll start enjoying what you do too much."

"She could be the one who provides the spark that could mean the difference between your next celebration being a birthday or a funeral."

Stone gave him Carly's number.

They began enacting Teddy and Carly's plan the next day. The first step was having Joan move the bugged clock from the garden shed to a bookcase in Stone's office.

Stone eyed it from his desk, still uncertain about having it there. Motion drew his attention to Teddy Fay, who was standing in the doorway, rolling his finger in the air, in a get-on-with-it gesture.

Stone nodded obediently, then activated the intercom, and said, "Joan, who's first on the call list?"

"Monica Anderson, Emerald River Shipping."

"Please get her on the line."

For the next few hours, he spoke to clients, making it sound like he was hard at work, while deftly avoiding confidential topics. Teddy disappeared for a few minutes now and then, but otherwise stayed just outside the room, observing.

A few minutes before noon, Joan buzzed him. "Dino on one."

Stone hit the speaker button. "Good morning."

"I'm checking to make sure you're still alive," Dino said.

"So far."

"Good to hear. If that changes, let me know. I'll need to send my black suit to the cleaners."

"Your concern is appreciated."

"Are you going to that dinner on Friday night? Or are you going to continue being a caveman?"

"Bill Eggers would never forgive me if I missed it. You were invited?"

"Not me, Viv. I'm going as her plus one. I assume Carly is yours."

"We are each other's."

"That's convenient."

"It is," Stone said. "Friday's still a few days away. I assume you'll be hungry before then."

"Past history says yes."

"You and Viv are welcome to have dinner here tonight."

"Great. Seven?"

"See you then."

Stone hung up, and then called Joan. "I'm going to grab lunch. I'll be back at two."

"I'll make sure things don't burn down in your absence."

Teddy Fay stepped out of the way so Stone could exit his office, and then followed him to the study. Carly was there already, waiting for them.

"How did it go?" she asked.

"Good, as far as I could tell," Stone said.

Teddy nodded. "Exactly as scripted."

"You're sure the bug picked everything up?" Stone asked.

"Each and every word." Teddy's occasional absences had been due to him checking equipment he'd set up in another room, to monitor what the bugs were recording and if they were transmitting properly.

Stone's cell phone rang with a call from Dino. He put it on speaker.

"Well?" Dino said.

"You're a natural," Teddy said.

"Maybe I should get a headshot and send it to a certain movie studio in Hollywood. I would hate for my talent to go unappreciated by the masses."

"I was speaking of your voice. As they say, you have a face for radio."

"He could be a character actor," Carly said.

"Don't encourage him," Stone said.

"Call me if you require an encore performance," Dino said and then hung up.

"So, what now?" Carly asked.

"Now, we eat lunch," Teddy said. "Then it will be your moment to shine."

CHAPTER 63

Sarge entered the suite at the Four Seasons that was serving as his mission headquarters and headed straight to where Deacon, his sound expert, was set up.

"I got a message that the bugs are working," Sarge said.

Deacon removed his headphones and smiled. "Oh, yeah."

"Since when?"

"Around nine AM. I didn't get around to checking until right before I called you." Deacon's excitement slipped a little. "Sorry. It's just we hadn't—"

"Don't worry about it."

Since the conversation they'd heard right after the clock had been delivered days earlier, the bug had picked up nothing else. The same was true of the listening devices in the wine basket.

"Do you know why it's working now?" Sarge asked.

Deacon shrugged. "I guess they got around to finding a place for the clock."

That had to be it. The Sarge had used the bugs before against more formidable opponents and never once had they been detected. No way an uptown lawyer like Barrington would find them.

"What have you heard?"

"Mostly work calls. Law stuff. But there is one conversation I know you'll be interested in."

Sarge leaned over Deacon's shoulder. "Play it."

Deacon hit a few buttons, and Barrington's voice came over the speaker. He was talking with his friend, the police commissioner. Sarge cocked his head at the mention of an event on Friday night that both men would be attending.

"Play it again," he said.

Deacon did.

The lawyer was finally going to leave his house. This was the break the Sarge had been waiting for.

"This Bill Eggers," he said. "I've heard that name somewhere."

"He's one of the partners at Barrington's law firm."

"You didn't hear anything else about the dinner?"

"No. Barrington went to lunch after that. I haven't heard anything since then."

Sarge took a step back, intending to go into the bedroom he was using, to make some calls. "Okay. Let me know if you—"

Deacon suddenly pressed his headphones against his ears. He then hit a button, and Barrington's voice came out of the speaker again.

"Look the file over and let me know if you have any questions."

"Will do," a woman's voice said.

"Is this live?" the Sarge asked.

Deacon nodded.

"Carly," Barrington said. "Still on for Friday, right?"

"I'm looking forward to it," the woman said.

"That makes one of us. You know how I hate these bar association dinners."

"It'll be fun. You'll be with me, remember."

"That is the saving grace."

"What time do I need to be ready?"

"Fred tells me we'll need to leave here at six-twenty to make it in time for cocktails."

If the woman said anything else, the bug didn't catch it. But that was fine. Sarge had heard more than enough. A bar association dinner on Friday night. The details would be easily obtained.

He clapped Deacon on the shoulder, said "Good work," and then headed into the bedroom.

Teddy gave Carly a thumbs-up as she walked out of the room, and she took an exaggerated bow. As she walked off, he signaled to Stone he was leaving.

He had work of his own to do.

Stone handed Dino a glass of Knob Creek as his friend entered the study that evening. "Where's Viv?"

"Outside, having a word with the head of your security team." Dino took a seat. "Where's Carly?"

"Here I am," Carly said, stepping into the room.

Stone handed her a drink.

"Thanks," she said. "I have a question. Am I reporting to you now or is it still Herb Fisher?"

"Officially, Herb is still your boss. Why do you ask?"

"I had a call from a potential client. Well, new client now."

"Who's the client?"

"Cabrera Cosmetics. Have you heard of them?"

"No, but that's not my area of expertise."

"They had a million and a half in legal fees last year, and their business has grown twenty percent in the last six months, so that amount will go up."

"What you're saying is that you're making it rain again."

"I guess I am."

Stone laughed. "Tell Herb, but I want to be there when you do to see his reaction."

"Why? Won't he be happy?"

"That will be one of his emotions."

"Is Billy not joining us?" Dino asked. "Or did he leave town already?"

"He's still in the city," Stone said, "but exactly where and as who, I don't know."

Dino held up a hand. "That's probably more than I need to know already. Forget I asked."

Viv walked in. "Forget you asked what?"

"I don't know," Dino said. "I've already forgotten."

"Which reminds me," Stone said, "thanks again for your assistance this afternoon."

"What assistance?" Dino said, innocently. "On a completely unrelated note, I thought you'd like to know there will be a substantial yet discreet police presence at the bar association dinner on Friday night. Coincidentally, this includes the route from your house to the hotel."

"How fortuitous," Stone said.

"And you'll have Strategic Services escort vehicles from here, and a team of bodyguards at the site," Viv said.

"Maybe with all that," Carly said, "the Sarge won't even try anything."

"I wouldn't count on it," Viv said. "I've become a Sarge expert over the last few days, and from what I've learned, he doesn't like to pass up an opportunity."

Dino stood. "Anyone else need another drink?"

Stone downed what was left in his glass and held it out. "I do."

CHAPTER 64

"Can I help you?"

The new arrival stepped inside the dingy office, located in a distribution center that looked as if it had been mothballed years ago. He approached the beat-up metal desk the guy who'd spoken sat behind. "I'm looking for someone called the Corporal."

"Is that right? Who are you?"

"Why? Are you the Corporal?"

"If you don't want to answer my question, you can just turn around and leave."

"Hey, no need to get that way. I was just asking, that's all. My friends call me Dial."

"Who said we were friends?"

"Fine," Dial said. "On my birth certificate, it says Henry Sommers. Is that better?"

The guy looked him up and down. "Why do people call you Dial?"

"Because when I'm working, I'm dialed in. You know, focused."

The guy snorted and rolled his eyes. "Who sent you, *Dial*?"

"I got a call from a friend. Said you were looking for people."

"What friend?"

"Dieter Mainz."

The guy's eyes narrowed. "*You* know Dieter Mainz?"

"I just said I did, didn't I?"

"Hold on."

The man disappeared into another room, closing the door behind him.

Henry "Dial" Sommers—aka Billy Barnett, aka Teddy Fay—took a seat on the only other chair in the room, and rolled his head over his shoulders, like he didn't have a care in the world.

The truth was he did not know Dieter Mainz personally, but he knew someone who knew the German mercenary. That someone had made it clear to Mainz that his cooperation would go a long way in keeping him from being turned over to some very bad people he had pissed off.

Ten minutes later, the door opened again, and the man returned. "Come with me."

He led Teddy through the building and into a large garage. Half a dozen men were hanging around the far corner, not talking to each other and looking like they were waiting for something. Two other men were standing next to a black Suburban that Teddy would bet was armored.

"Wait here," his escort said, then jogged over to the two at the SUV.

A few seconds later, he waved for Teddy to join them.

When Teddy did, the guy said to the others, "This is Dial," without his previous sarcasm.

"Thanks, Sammy," the older of the two said. "That'll be all."

The escort—Sammy—nodded and left.

"I'm the Corporal," the older one said. He didn't introduce his friend. "I talked to Dieter. He said you're good in a tight spot. Trustworthy."

"Happy he thinks so."

"Says you know how to shoot, too."

Teddy nodded.

"I understand you did some contract work for the Agency?"

"Among others."

"How long ago was that?"

A shrug. "A few years."

"Why did you stop?"

"Did my job too well."

"Dieter mentioned an incident in Rome."

"That's the one."

The Corporal looked at him, waiting for more, and Teddy looked back, making it clear that was all he was going to say.

"Dieter also said you didn't talk much."

"Talking is overrated."

The Corporal huffed then studied him again. "I like you, Dial. The job runs through Friday night. Ten K, flat fee. You interested?"

"What type of job is it?"

"Termination."

"Opposition?"

"Potentially considerable."

"You supply the hardware?"

"Yes. If you have special requests, we can try to accommodate that, too, but no promises."

Teddy acted like he was considering the offer. "All right, I'm interested. When do I start?"

"Right now." The Corporal held out his hand and Teddy shook it. "Welcome to the team."

CHAPTER 65

Asimov received a call from the Sarge on Thursday afternoon.

"I'll bring Barrington to you by midnight tomorrow night," Sarge told him.

"What's the plan?"

"That's not important. All you need to know is that his part of our contract will soon be complete."

Perhaps Asimov didn't *need* to know, but he wanted to. "You can't tell me anything?"

"If you don't know, it will be easier for you to deny involvement."

Though true, Asimov didn't like the way the Sarge had said it, as if he were teaching Asimov a lesson. "And Rawls? Will he be there, too?"

"He is still in London."

"Then when will he be dead?"

"When I tell you he will be."

That was one step too far. Asimov opened his mouth to put the Sarge in his place, then realized the Sarge had hung up.

That son of a bitch.

Asimov took a deep breath. Barrington would soon be dead; that's what he needed to concentrate on.

The timing could not have been better. A few members of the council who had voted for him had begun to grumble about how long everything was taking. Soon, he knew, they would start having second thoughts about his leadership.

He called his secretary. "Inform everyone we will have a council meeting at nine o'clock tonight, at the usual location."

"Yes, sir."

This time, Asimov waited until the others had all arrived and were seated before he entered the room. As instructed, as soon as he took the chair at the head of the table, four women entered. They were all in their early twenties, beautiful, and wearing dresses that showed off more skin than they covered. Each carried a bottle of wine, and made their way around the table, serving the members of the council, then filed out of the room.

Once they were gone, all eyes turned to Asimov.

He lifted his glass. "A toast."

The others raised their flutes.

"To the family," he said

"To the family," the rest of the council repeated.

After everyone had taken a drink, Asimov said, "I have news. As we all know, the lawyer Stone Barrington has been a blight on our family for some time now. But that ends tomorrow night. Before the sun comes up on Saturday, he will be dead."

"Barrington has escaped our attempts before," the man who oversaw the family's illegal gambling interests said. "Are you sure he will not do so again?"

"I have taken measures to ensure that doesn't happen."

"Anton Pentkovsky thought the same. As did the Gromykos."

Several of the men around the table nodded in agreement.

Asimov's jaw tensed. He had been expecting a more positive reaction. "I am neither Anton Pentkovsky nor either of the Greeks. If I say it will be different this time, I mean it."

The Bean Counter leaned forward and gave Asimov a sympathetic smile. "Perhaps sharing your plan with us will help everyone feel the same confidence."

"Yes," someone else said. "We would like to know."

Voices of assent rose in support.

"I can't give you the specifics," Asimov said, "but I can tell you that I have hired the best in the business to do the job. I'm sure you have heard of the Sarge?"

The looks of skepticism he'd been given quickly turned into one of disbelief.

"The Sarge?" one of the men said. "That was a smart move."

"Thank you," Asimov said.

"He has an excellent reputation," another offered.

"And the Sarge is taking care of Barrington tomorrow?" a third asked.

"He is," Asimov said.

Smiles began breaking out. If the Sarge was on the job, then there was no way it could fail.

The Bean Counter rose and raised his glass toward Asimov, and said, "Then I think another toast is in order. To the imminent demise of Stone Barrington." He drank.

The others did the same.

Asimov smirked, and then motioned to the man guarding the door. "More wine!"

Lauren was at her desk when the Bean Counter returned with Korolev, who had once again been his driver. The Bean Counter motioned for both for them to come into his office.

Lauren shot Korolev a questioning look. He shrugged and they followed their boss inside.

"Close the door," the Bean Counter said.

Lauren did so.

The Bean Counter sat behind his desk and pointed at

the chairs on the other side. "Sit." When they had, he said, "You both have a choice to make."

"Choice?" Korolev said.

"A simple one. You need to decide where your alliances lay. And I'm not referring to the two of you sleeping together."

Both Korolev and Lauren stared at him, wide-eyed.

"As long as it doesn't affect your work, I don't care," the Bean Counter said. "What I'm asking is if your loyalty is to me or Dmitri Asimov." He looked directly at Lauren. "I know you were devoted to Peter Greco, and that you kept him informed on what was happening in the family when he was pulling back."

She gulped. "I—I—"

"I don't care. I let you do that because it was useful to me. But Greco is gone."

She stammered again, but no words came out.

Korolev said, "Mr. Asimov is the head of the family."

"Is that your answer, then?"

"You," Lauren said. "I choose you." She looked at Korolev, urging him to say the same.

"I have sworn my loyalty to the family," Korolev said to the Bean Counter. "I've had few dealings with Mr. Asimov, but those I have had have not been . . . ideal. *You* have never done wrong by me."

"I need you to say it," the Bean Counter said.

"My loyalty is with you," Korolev said, after a moment. "Why are you asking us this?"

"Because there is an excellent chance that Asimov will be out of favor in the next twenty-four hours. The family is already fragile. The choices made in the wake of his screwup will be the difference between whether the family survives or not. With your help, I will bring stability to the organization."

"What do you need us to do?" Korolev said.

CHAPTER 66

Stone woke up on Friday morning to the touch of Carly's hand bringing him to attention. He returned the gesture until she could stand it no more and rolled him onto his back to show him how much she appreciated his efforts.

Later, as they lay blissfully beside each other, Stone said, "Is it just my imagination, or was that particularly enjoyable?"

"I hope so," Carly said. "I put in extra effort."

He turned on his side to look at her. "Are you saying you don't usually put in the effort?"

"Not at all. I always put in the effort. This was just extra." She swung off the bed and stood, giving him a view he appreciated.

"I'm not ungrateful," he said, "in fact, I'm just the opposite, but is there a reason why?"

She stopped in the doorway to the bathroom, looked back, and said matter-of-factly, "If things don't go well tonight, this could be our last time."

She disappeared through the doorway, and Stone rolled onto his back, feeling suddenly less relaxed than he had a moment before.

Yesterday, he had received a short message from Teddy Fay, confirming that their plan had worked, and he was indeed going to be targeted at some point Friday evening. What Teddy hadn't said was when the attempt would be made. The only other thing in Teddy's message was: *Remind Carly that she should be armed. You, too, if you can bring yourself to do it.*

Carly walked out of the bathroom twenty minutes later, a towel wrapped around her hair and nowhere else. She tilted her head when she saw him. "What's wrong? You look unhappy."

"I don't know. Perhaps I wasn't expecting the woman I'd just slept with to remind me that I might die tonight."

"Oh, should I not have said that?"

"I could have gone without hearing it, especially in the way it was presented."

"I'll remember that for next time."

"The next time I might die?"

"In my experience, it's not a rare thing."

He couldn't argue her point.

Later that morning, Joan appeared in the doorway of his office, holding up a piece of paper on which was written: *Dino on 1.* The bugged clock was still on his bookshelf, so caution needed to be taken in its presence.

He relocated to the room Carly was using as her office, where she was engrossed in something on her computer. "I need to use your phone," he said.

Without looking at him, she waved at it. "Help yourself."

He picked up the receiver and punched line one. "Hi, Dino."

"Thought you'd like to know, I had the bomb experts sweep the hotel, and they didn't find anything."

"Strike explosions off the list."

"Don't be too hasty. He could always fire an RPG at your car."

"Are you purposely trying to up my anxiety level, or is there some other reason you called?"

"I wanted to remind you to make sure you're carrying this evening. It's not a night to forget that kind of thing."

"So I've been told by other sources."

"Other sources would be correct," Dino said.

"I'll see you tonight."

"Here's hoping."

Stone hung up and noticed Carly was still staring at her computer. He leaned over so he could see the screen. On it was the website for the CIA.

"Carly, did Lance Cabot contact you?"

"No. Was he meant to?"

"Quite the contrary. He promised me he wouldn't."

"Then he's living up to his word."

"Why are you on the CIA website?"

"Research."

"What kind of research?"

"You *did* say Lance would talk to me eventually. I want to be prepared for when that happens."

"That site will only tell you one side of the story, and not a complete one at that."

"Which is why I have also read four books and twenty-seven magazine articles, from differing viewpoints."

"Carly, you are not seriously considering accepting an offer from Lance, are you?"

"No offer has been given, so there is nothing to consider yet."

"When it is given, what are you supposed to do?"

"Say no, and then talk to you."

"Please, do not forget that."

"My brain wouldn't let me, even if I wanted to."

CHAPTER 67

Stone checked himself in his mirror. Dinner was a black-tie affair, so he was wearing a Tom Ford tuxedo.

"Wow," Carly said from behind him.

He turned. She had just exited her dressing room wearing a dark blue floor-length dress, with a neckline that plunged past her chest and a slit that ran up her right leg to her hip.

"Wow, yourself," he said.

"You like it?" She slowly twirled around. "It's a Ramy Brook. I realized I had nothing I wanted to wear, so I picked it up at Bloomie's yesterday."

"I approve, but only if you took Fred with you."

"Even if I had tried to go alone, he wouldn't have let me."

"Remind me to give him a little something extra in his paycheck this month."

There was a knock on the bedroom door.

"Come in," Stone said.

Helene entered. "The man from Strategic Services is downstairs."

"Thank you, Helene." He held out his arm to Carly. "Shall we?"

Waiting for them at the base of the stairs was a man who was at least six and a half feet tall and had the body of a linebacker.

"Mr. Barrington, I'm Richard Ray. I will be in charge of your personal detail tonight."

Stone shook hands with him. "Nice to meet you, Richard. This is Carly Riggs, a fellow attorney at Woodman & Weld, and my date tonight."

They exchanged greetings.

"Your Bentley and driver are in the garage, along with one of the vehicles that will be escorting you. I will be riding in the latter. I've been briefed on the specs of your car, so I know how difficult it would be to breach, but that does not mean someone won't try."

"It's happened before," Stone agreed.

"So I've heard. My goal is to make it as unattractive as possible for anyone to do so tonight. To that end, there will be three vehicles ahead of us, reconning upcoming intersections. We will not pass through any unless we receive an all clear first. Also, we will be coordinating with the police units set up along your route. If they spot something suspicious, we will immediately return here."

"It sounds like you have all the bases covered."

"No one ever has all bases covered, but hopefully we will have done enough to dissuade the enemy."

"I have every faith in you."

"Thank you. One last thing."

"Yes?"

"Mr. Freeman wanted me to ask: Are you carrying?"

Stone patted the bulge under his jacket. "Yes, and I prefer it to stay where it is."

Fred waited for them in the garage, along with a pair of Strategic Services bodyguards in a sedan.

Once everyone was in their respective vehicle, Fred raised a walkie-talkie to his mouth. "This is Fred Flicker. Ready when you are."

Ray's voice came over the speaker. "Proceed, Mr. Flicker."

They drove out of the garage. Waiting on the street were five additional Strategic Services sedans. Three raced ahead to check the intersections. One of the remaining pair pulled in front of Fred to lead the way, and the other fell into place at the back of the motorcade.

"This is exciting," Carly said.

Stone looked at her.

"What?" she asked.

Stone shook his head. "Nothing."

The radio crackled to life. "Barrington's car just left his garage. There are six vehicles traveling with it. Looks like half of those are advance scouts, checking the route."

Sarge picked up his mic. "Can you confirm Barrington is inside?"

"Not yet, but they'll be passing our position in thirty seconds. I should be able to tell then."

Sarge looked out the window of the van. It was parked in an alley, three blocks from the hotel where the bar association dinner was to take place.

He had four watchers stationed along the presumed route Barrington would take, to track the lawyer's progress. The rest were waiting at their assigned locations.

Even before Sarge received word through Asimov that the police would be watching the lawyer's route to the event, he'd discarded the idea of going after Barrington on the streets. There were too many factors the Sarge couldn't control to make it viable, if he were to honor Asimov's request to see Barrington before he was killed. Chief among them was Barrington's armored vehicle.

The voice of the spotter came over the radio again, "They just passed. Barrington's in the back seat with the same woman we've seen going into and out of his place. His normal driver is behind the wheel."

"Copy," Sarge said. He switched the radio to the frequency setting used by the rest of the team. "This is the Sarge. The target is en route."

Immediately after the Sarge's announcement, one of the guys in the back seat of the Ford crew cab pickup parked behind the van said, "I guess this is really happening."

The Corporal, who sat in the driver's seat, shot him a

look in the rearview mirror and said, "Can it." He then nodded at the man in the front passenger seat, the guy called Dial.

Teddy Fay picked up the microphone and said, "Team two, copy."

CHAPTER 68

The motorcade made it to the hotel without trouble. As Fred pulled the Bentley into the hotel garage, Ray's voice came over the radio. "The elevators are just ahead. Stop when you reach them. But no one get out until I give you the okay."

Fred stopped where he'd been told, and the sedans escorting them did the same. Several Strategic Services men piled out of each vehicle. Half took up positions around the Bentley, while the other half spread out through the area.

Finally, Ray knocked on the window beside Stone, signaling for Fred to disengage the locks, which he did.

When Stone and Carly exited the Bentley, four bodyguards fell in around them.

"This way," Ray said and led the entourage to the elevators.

It was a tight fit, and Stone could feel Carly fidgeting beside him. "Are you okay?" he asked.

"Not a fan of crowded elevators," she said.

"Apologies," Ray said. "We'll be there soon."

"Be thankful you're not Holly Barker," Stone said. "It's like this for her all the time."

"I can't be president until I'm thirty-five, so I still have a few years to get used to it."

"Let no one ever say you lack confidence."

Soon, the door opened on the seventh floor and revealed four more bodyguards standing outside.

"This might be overkill," Stone said to Ray.

"Mr. Freeman thought you might say that. He told me to tell you, 'better overkill than being killed.'" Ray guided them through a service corridor to an unmarked door. "The dining room is through here. You are sitting at table number one. It's center front, near the stage."

"You're not coming in with us?" Carly asked.

"No, ma'am," Ray said. "We have several people inside already. When you want to leave, let one of them know, and I will meet you here with my team." He opened the door. "Have a wonderful evening."

A server made his way around the ballroom with a bottle of wine, topping off the glasses of the attendees who had arrived early, and filling those for the guests who had just found their seats.

"More sauvignon blanc?" he asked an elderly woman sitting at table eleven.

"Please," she said.

He filled her glass, her third already. As he moved to the next guest, he caught movement in the corner of the room. A man and a woman had just entered the ballroom, dressed respectively in a tux and a blue dress with a slit that showed an impressive amount of thigh. Though they were obviously guests, they had come in through a doorway that connected to a staff-only area, instead of using one of the main entrances.

The server moved to the next table, where he filled another glass before stealing a peek at the new arrivals. This time he was sure of who they were.

"Can I get some of that?" a man a few seats away asked.

The server smiled sympathetically. "Of course, but I need to get a new bottle first."

The man pointed at the one in his hand. "It looks like there's enough in there for another glass."

The server smiled again. "I'll be right back."

He walked away before the guest could say anything else, leaving the ballroom through the first door he came to. He checked to make sure no one was watching him, then pulled out his phone and sent a text that read: **Target in ballroom.**

The radio in the truck came back to life with the sound of the Sarge's voice. "All teams to your positions."

The Corporal started the engine, then followed the van as it began to move.

"Check your gear," the Corporal ordered.

Teddy removed his Sig Sauer P320 Nitron 9mm from his shoulder holster, checked the magazine and chamber, then attached the silencer, and set the weapon in his lap.

If the Corporal had been less secretive about the scale of the operation, Teddy could have ended the threat before they'd gotten to this point. But missing a vital piece of the puzzle could have resulted in Stone losing his life. That would be mission fail for Teddy.

The van and truck entered the alley behind the hotel where the banquet was being held.

"Come in, team three," the Sarge said over the radio.

"Go for team three."

"Status?"

"Everyone is in the ballroom. Ready when you are."

"Team four," the Sarge said.

"Go for team four."

"Start the countdown."

"Copy. Starting countdown. T-minus ninety seconds."

The van stopped next to a loading dock at the back of the hotel, and the truck pulled in behind it.

"Sixty seconds," the man from team four announced.

"Radios on, night vision ready," the Corporal said, "and everyone out."

Teddy donned his earbud and night vision goggles, tilting the latter up so they were off his eyes, then he exited the truck with the others. More men climbed out of the van, including the Sarge.

This was the closest Teddy had been able to get to him so far, but he still wasn't close enough.

The Sarge motioned everyone to a door on the dock, and they all jogged into position.

Over the radio came, “Fifteen seconds.”

The Sarge looked around. “No mistakes.”

The men answered with nods.

“Five . . . four . . . three . . . two . . . one . . .”

CHAPTER 69

In the ballroom, Bill Eggers and Herb Fisher smiled as Stone and Carly walked up to table number one.

"Stone," Bill said, "we were starting to wonder if we would ever see you again."

"What?" Stone said. "And deprive you of my charm? Never." He shook Bill's hand. "Congratulations on being honored tonight."

"I think it was the only way they could get me to come to the dinner," Bill said.

"You and me both."

"Count me in on that," Herb Fisher said.

Bill turned to Carly. "And here's our rising star. Good evening, Carly."

"Good evening."

"Carly, don't you have something to tell Herb?" Stone said.

Bill's brow creased. "No one's tried to poach you away from us already, have they?"

"Not officially."

Her answer did not assuage Bill's concern.

"Lance Cabot made an inquiry," Stone said, "but I told him in no uncertain terms she was not available."

"It's my understanding he can be tenacious," Bill said.

"He can be, but so can I. If he does talk to her, Carly and I have already agreed to have a very serious conversation about what a job with Lance would mean." Stone smiled at Carly. "Isn't that right?"

"Tell Lance no and talk to you," she said.

"See," Stone said to Bill, "nothing to worry about."

Bill looked appeased.

"You haven't said what you wanted to tell me yet," Herb said.

"Oh, right," Carly said. "I brought in another client. Cabrera Cosmetics."

"One point five million in billing last year," Stone said. "And they expect that to go up."

"That's fantastic," Bill said. "At this rate, you'll be a partner in a year."

Carly looked at Stone. "See. That's what I—"

Stone held a finger to his lips, stopping her, and tapped his temple to remind her that was one of those things to keep to herself.

"Right," she said. "Sorry."

"Congratulations, Carly," Herb said. "You know, Stone, you seem to have already taken her under your wing.

Maybe we should make it official, and make you her supervisor."

"If I did that," Stone said, "I'd miss out on watching your misery."

They sat down, and a server approached.

"Some sauvignon blanc, sir?" the man asked Stone. "Or, if you prefer, my colleague has pinot noir."

"Sauvignon blanc will be fine, thank you."

The server poured, then asked the same of Carly.

"I think I'd like to try the pinot."

As the last word left her mouth, the room plunged into darkness. Even the emergency lights that should have come on remained unlit. At that same instant, Carly heard the sound of something familiar, but didn't immediately place it.

Someone grunted and she felt movement next to her. "Stone?" She put a hand where she thought he was, but his chair was empty. "Stone?"

There was no answer. What she did hear was a cacophony of confused voices and the sound of several feet moving across the room.

Here and there flashlights on phones started coming on. Carly pulled her mobile from her clutch and turned on its light. Stone was nowhere to be seen.

She looked around for the Strategic Services bodyguards but didn't see them anywhere.

"My God!" a woman yelled from the side of the room. "Sir, sir, are you okay?"

Carly swiveled in her chair toward the voice. The woman was crouched over the silhouette of a man on the ground, lying in the exact spot one of the bodyguards had been standing.

Someone else ran up to the woman. "Is he all right?"

"He has a pulse but he's unconscious."

Carly quickly glanced to the other places she'd seen Strategic Services personnel and spotted more silhouettes on the ground.

That's when the identity of the noise she'd heard clicked into place. It had been a spraying sound. The same sound she had heard right before she'd been kidnapped.

She reached under her dress and withdrew the Smith & Wesson Equalizer pistol from her thigh holster and ran.

CHAPTER 70

A voice said, "We have the package."

To Stone, it sounded like whoever was speaking was right next to him. Which was odd enough, but even stranger, Stone seemed to be moving, but not by his own power. And then there was the fact that something hard was pressing against his chest.

It took him two tries to open his eyes without feeling dizzy.

He appeared to be in a giant empty room, lit by the moving beams of several flashlights. Only the floor was above him, passing quickly under a pair of feet that couldn't be his. He could hear other footsteps, too, so there had to be multiple people around him.

When he realized his cheek was bouncing off someone's torso, it all came back to him. He'd been sitting at the table in the ballroom, talking with Bill Eggers, Herb Fisher, and Carly, when all the lights had gone out. Then some kind of liquid spray had hit him in the head, landing more on his ear than his face, and the next thing he knew, he was here, in what was clearly another, albeit unused, ballroom.

Recalling the spray triggered another memory, distant and foggy. He concentrated and then it came to him. The same knockout spray that had been used on Carly when she'd been kidnapped must have been used on him, only the person doing the spraying's aim had been off. If Stone had been hit square in the face, no telling how long he would have been unconscious.

"Get the door," the man carrying him said.

A pair of footsteps ran ahead, and moments later, a door opened.

Stone had no idea where the door led, but he was sure he couldn't let himself be taken through it.

Marshaling as much strength as he could, he swung his elbow upward into the underside of the man's jaw.

His kidnapper's head snapped backward, and the guy yelled in pain.

Stone had expected that much, but he hadn't expected the crack of a gun that echoed across the room at the exact same moment his blow had landed. The boom was quickly followed by the sound of the door that had opened closing again.

Ignoring both, Stone rolled off the man's shoulder and fell toward the floor, where he landed awkwardly on his hands and feet, almost but not quite like a cat.

A second bang ripped through the air, and one of the people behind Stone dropped lifeless to the floor, the person's flashlight smacking against the tile, and spinning around and around until coming to rest, pointing at a wall.

"Stone, stay down!" Carly called out.

A third gunshot was accompanied by a flash that briefly lit up Carly's position, and another body dropped.

The guy who'd carried Stone had recovered enough to rush to a nearby door. He shoved it open and disappeared through it.

Stone reached for his gun, but it wasn't there. He'd finally followed Dino's advice to carry, and someone had taken his weapon. Just his luck.

He started crawling toward one of the people Carly had shot to borrow their weapon when a hand grabbed the collar of his jacket and yanked him back.

"Where the fuck do you think you're going?"

Stone twisted out of his jacket, sending the guy who'd grabbed him stumbling backward, and scrambled to the nearest downed man.

His hand landed on the butt of a pistol just as Carly yelled, "Stone! Watch out!"

He whirled around, his fingers gripping the gun. In the dim halo of a flashlight, he saw the guy from a moment before barreling toward him.

Stone raised the pistol and fired; the sound was dampened by the attached silencer.

The man staggered backward as a dark spot grew on

his chest. He looked at his shirt, confused, then collapsed two feet away from Stone, unmoving.

Stone heard feet running toward him and swung the gun toward the sound.

"It's me," Carly said.

Stone lowered his weapon.

"Did we get them all?" she asked.

"No," he said, nodding at the nearby door, "one of them went through there. Let's get out of here before he comes back."

As Stone pushed himself to his feet, the door at the far end of the ballroom banged open.

Dial, aka Teddy Fay, had been assigned guard duty on the hotel service entrance, with three other men. Two were stationed outside, while he and the remaining guard were to keep watch inside. As soon as they'd taken their positions and his partner looked the other way, Teddy shot him in the head, then stepped into the doorway and did the same to the other two.

He whirled around and hurried in the direction the Sarge and the other men had gone, soon spotting two more of the Sarge's men guarding a stairwell door. He dropped them with head shots before they knew he was there.

As he stepped into the stairwell, muffled gunshots echoed down from above.

Teddy raced up the stairs.

Korolev had watched the Sarge and his people enter the hotel from atop a building on the other side of the alley, and then witnessed one of the Sarge's men come back out and shoot the two who were guarding the door.

What the hell?

He made his way down to the alley and sneaked up to the hotel service entrance. It only took one glance to confirm both lookouts were dead.

Through the partially opened doorway, he heard several distant gunshots, and immediately retreated down the alley and called the Bean Counter.

"I don't know what the Sarge's plan is," he told his boss, "but I don't think a gunfight inside the hotel is a part of it."

"Come get me," the Bean Counter said. "It's time to clean up this mess."

CHAPTER 71

With no time to hide anywhere else, Stone grabbed Carly by the arm and pulled her to the doorway the man who'd been carrying him had disappeared through.

"Stay low," he said and pushed the door open.

He expected to find a corridor or another ballroom on the other side, instead they rushed onto an outdoor deck that served as a restaurant dining area. Though it, too, was affected by the blackout, there was enough residual city light for Stone to make out the darkened forms of tables and chairs and a bar. There were also people, too; at least two dozen. Some had phone flashlights on, and all were murmuring excitedly about the loss of power.

"Who's there?" a voice called from several feet away. Stone instantly recognized it as that of the man who'd been carrying him.

Stone took Carly's hand and quietly led her away from the door, in the opposite direction from where the voice had come. They found a row of wide urns filled with plants and slipped behind them.

"Adams, is that you?" the man called.

"I'm sorry," one of the nearby diners said. "Are you talking to me?"

Ignoring him, Sarge's man said, "Whoever you are, I know you're here."

"Hey, this blackout is scary enough without you talking like that," the diner said.

"Shut up or I'll show you how scary I can be."

The diner shot up out of his chair, and he and his companion moved farther into the dining area.

A flashlight flicked on in approximately the same place the mercenary's voice was coming from. The beam swung to the door then across the floor toward the urns, tracking the path Stone and Carly had taken.

Stone stuck his gun between two urns, aimed at the light, and pulled his trigger.

Two things happened simultaneously. The flashlight beam twisted wildly, before whipping around and pointing back Stone and Carly's way. And though Stone's gun had a silencer, the sound of the shot was still loud enough to cause several people in the restaurant to scream. This, in turn, motivated the rest of the diners and the staff to rush toward the main exit, knocking over tables and chairs as they went.

"Nice try, asshole," the mercenary said over the cacophony, clearly uninjured.

His gun fired, and an urn just to Stone and Carly's left exploded into a million pieces.

As Stone raised his pistol to try again, Carly whispered, "You're a terrible shot. Let me."

Before he could reply, she fired, her unsilenced weapon booming across the deck.

The man grunted, and the flashlight beam arced upward and then went out as both it and the man crashed to the ground.

"Not bad," Stone said.

"It's really not that hard."

"I'll take your word for it."

Stone had expected to hear more screaming, but it appeared everyone had made their escape. Which was exactly what he and Carly needed to do.

"Come on," Stone said.

He grabbed Carly's hand again, and they moved into the dining area, weaving through the tables toward the exit.

A moment after the ballroom door was pulled open, Sarge and his men heard another door across the room open and close, and after that silence.

"Team three!" Sarge called into the room.

Nothing.

He tapped the shoulder of the man nearest the door. "Check."

The man dropped to a crouch and peeked through the doorway. "Oh, shit," he said.

"What?" Sarge whispered.

"All I see are bodies."

"Everyone in," Sarge ordered.

The first two entered, their guns raised. The next two followed, doing the same, and then the Sarge and the Corporal went in.

"What the hell?" someone said.

There were several bodies on the floor, all within a few feet of each other.

"Go," Sarge said.

The men moved cautiously toward the bodies, sweeping the area with the barrels of their guns. When they reached them, they checked each for a pulse. All were dead.

That was bad, but more troubling to the Sarge was that none of them were Barrington.

"What happened?" the Corporal said.

Through gritted teeth, the Sarge said, "A clusterfu—"

A gunshot drowned him out.

Everyone ducked and looked around.

"Where did that come from?" someone asked.

The shot had been loud, so it had to have been close, and it wouldn't have been caused by one of his men. They were all carrying silenced weapons. But there was no sign of a shooter.

A second shot rang out.

"There," the Corporal said, pointing at a doorway near where Sarge and the others were.

Everyone ran toward it.

Stone and Carly were just over halfway across the deck when the ballroom door flew open, and several silhouettes rushed out and took cover. Stone motioned to the bar as it was the closest place to hide.

As they crept behind it, they heard one of the new arrivals say, "There's another body. It's Craddock. He's dead."

"See," Carly whispered to Stone. "Not hard."

"Can we discuss your marksmanship later?"

They peeked around the edge of the bar at the new arrivals.

"I count five," Carly whispered into Stone's ear.

Stone was about to agree when two more silhouettes exited the ballroom.

"I mean seven," she said.

"Barrington!" an angry voice yelled. "I know you're here. Come out. You're only prolonging the inevitable!"

"He sounds upset," Carly said.

"I'm not feeling particularly happy at the moment myself," Stone said.

"Do you think that's the Sarge?"

Across the room, someone said, "Sarge, I found something."

"What is it?" the angry voice said.

"Yeah, I think it's the Sarge," Stone whispered to Carly.

"It's a woman's shoe," the guy who'd found something said. "A nice one."

"Sorry," Carly said. "I took them off when I first stopped to shoot, then dropped them when we came out here."

"Barrington," the Sarge yelled, "I know you're not alone. I tell you what. You give yourself up, and I'll let your companion go unharmed."

Stone peeked around the bar hoping to make out at least a shadowy shape, but the others were now hidden from view.

"I hear your mind turning," Carly said.

"I'll try to think quieter," Stone said.

"If you're considering taking him up on his offer—"

"I'm not. He's bluffing."

"You took the words out of my mouth."

"That's a first."

The Sarge said, "Everyone, spread out."

"Oh, good," Stone said under his breath. "For a moment there, I thought we were in trouble."

"I think we *are* in trouble," Carly whispered.

"I was quoting Butch Cassidy. It's what he says to Sundance at the end of the movie."

"What movie?"

"*Butch Cassidy and the Sundance Kid.*"

"That's a movie?"

"It is, and a great one. Did I just find a hole in the mental database?"

"I think so," Carly said, annoyed.

"Barrington, I'll give you to the count of three," Sarge yelled.

"You can worry about it later," Stone whispered to Carly. "For now, just shoot at anything that moves."

"One," the Sarge bellowed. "Two."

Teddy Fay was halfway across the body-strewn ballroom when he heard the Sarge yell, "Barrington, I'll give you to the count of three."

Teddy reached the door as the Sarge yelled, "One," and pushed it open as Sarge yelled, "Two!"

There were five of the Sarge's men spread out in front of him, heading into a deserted restaurant. Closer, to Teddy's left, was the Corporal, and ten feet beyond him, the Sarge.

The Corporal had twisted around as the door opened, and had started to raise his gun, but lowered it when he realized it was someone he knew.

"Dial, what the hell are you doing here?" he asked.

Teddy answered by putting a bullet through the man's forehead. He sent another shot toward the Sarge, but the Sarge had excellent reflexes and dove for cover the second Teddy had raised his weapon.

The other men, whose backs had been to him, were not quite as quick. Teddy fired five more shots. Three of the men went down, two of them permanently, while a fourth staggered out of sight, his left shoulder useless. The fifth was the only one other than the Sarge to get away cleanly.

Teddy slid behind a drink station just as the Sarge opened fire on him. The bullets dinged off an expresso machine and ripped through the cabinet sending shards of wood and plastic raining down on Teddy's back.

As he scooted to the far end and peered around, he put a fresh magazine into his gun. He could see the man with the shoulder wound crouched behind a table that had been tipped on its side. The man was sweeping his gun back and forth, but he'd lost his goggles, so his vision was limited.

Teddy pulled his trigger, ending the man's misery.

Stone and Carly ducked down as the gunfight broke out at the other end of the deck.

When Stone realized no bullets were flying their way, he whispered, "Let's get out of here."

With Carly on his heels, he edged out from behind the bar in a crouch, intending to head straight to the exit. But he'd taken only one step in that direction when the lights flickered back on.

Stone squinted from the sudden brightness. At the same time, he heard a grunt coming from just to his right, and the clatter of something hard hitting the ground.

Stone had just enough time to realize it was a man dressed in all black before Carly shoved him onto his hands and knees. Above him, her gun roared.

The man in black spun against a table, then fell to the deck, next to the set of goggles he'd thrown off when the lights came on.

Stone looked at Carly, who was crouched over him, her gun still pointing at the now dead man.

"Thanks," he said.

"You're welcome."

Heavy footsteps approached from somewhere behind them. Stone looked past Carly just in time to see a large man sprint out from behind a table. From the angry snarl on his face, Stone had no doubt of the man's identity.

As the Sarge brought up his weapon, Stone shoved Carly out of the way and twisted his pistol around. With no time to aim, he pulled the trigger the moment he thought the barrel was pointed in the right direction.

The Sarge's snarl slipped, and his brow knit in confusion. A moment later, blood began flowing out of a wound on the side of his neck, while at the same time, a bloom of red appeared on his shirt in the center of his chest.

He dropped to his knees and tried to speak, but the only thing that came out of his mouth was more blood. He then fell onto the floor.

Stone was about to push to his feet when another person appeared near the Sarge. He was dressed in the same black outfit the Sarge was wearing, but his gun was pointed not at Stone and Carly but at the dead man.

He kicked the body, and when it didn't move, he looked at Stone. "Are you guys all right?" the man who didn't look like Teddy Fay asked in Teddy Fay's voice.

"I am," Stone said. "Carly?"

"Me, too," she said.

"Is there anyone else we have to worry about?" Stone asked.

Teddy nodded his chin at the Sarge. "He was the last."

"Is he dead?"

"Dead as they get."

Stone and Carly rose to their feet and joined Teddy.

"He's got two bullet wounds," Carly said, surprised.

"Very observant," Teddy said. "Stone, I think you hit him a second before I did. Excellent shooting, by the way. Center mass. That's what you'd call a bull's-eye."

"I didn't even aim," Stone said.

"Maybe you should try not aiming every time," Carly said.

"At least it wasn't a miss."

"Not even a near one," Teddy said. "Time for me to leave. This place is going to get very crowded, very fast."

He hurried away without another word.

CHAPTER 72

"The Bean Counter is here to see you," Asimov's secretary said over the intercom.

"Tell him I'm busy," Asimov said. He'd been waiting to hear from the Sarge, and until he did, he wasn't in the mood to talk to anyone else.

"He says it's urgent."

"I don't care what—"

The Bean Counter's voice came over the intercom. "It's about the Sarge."

Shit. "I see. Um, come in."

The Bean Counter entered with another man Asimov had seen around, named Korolev.

Acting like everything was fine, Asimov said, "Have a seat."

The Bean Counter sat across the desk from him, but Korolev remained standing.

"So, you said this was about the Sarge?"

"I did. More specifically, about his attempt on Stone Barrington at the bar association dinner this evening."

Asimov momentarily considered denying any knowledge, but he knew the Bean Counter would see through the lie. "What about it?"

"You haven't heard?"

Asimov kept his expression neutral, while all the other muscles in his body tensed. "Tell me what you're talking about, and I'll tell you if I know."

Right before they'd arrived, the Bean Counter had received an update on the situation at the hotel from a source at the police department. "The Sarge is dead."

"What?"

"So, you didn't know."

"Where are you getting this information?"

"Is that really important?"

"It is if the information is wrong."

"It's not."

"Okay, well. Thank you for telling me. If you'll excuse me, I have some things to take care of." Asimov stood to see them out, but the Bean Counter remained in his seat.

"I don't think you understand the situation," the Bean Counter said.

"What situation?"

"Your situation."

The Bean Counter nodded to Korolev, who reached under his jacket, pulled out a pistol with an attached silencer, and shot Asimov twice in the chest.

CHAPTER 73

Stone woke to the sound of his cell phone vibrating on his nightstand. Beside him, Carly groaned and turned away without fully rousing.

He glanced at the clock and saw that it was nearly ten AM. He seldom slept this late, but last night had been a long one, even with Dino's help to free him from the investigators.

Speaking of Dino. He pressed accept on his phone.

"Don't tell me I woke you," Dino said.

"I'm not sure if you know this," Stone said, "but I had a rather eventful evening."

"I was there, remember?"

"Not until after the fireworks."

"Traffic, what's a guy gonna do? Speaking of the fireworks, I was given a message to pass on to you."

"From whom?"

"Gregor Dryga."

"Never heard of him."

"He's also known as the Bean Counter."

"That name I know. What's the message?"

"That for the foreseeable future the Russians have no interest in seeing you dead."

"That is not news I was expecting to hear this morning, welcome as it is. But is he even in a position to tell me that?"

"From my understanding, he is. He is now the head of the family."

"What happened to Asimov?"

"I was told his services were no longer needed, and that he has been retired."

"Permanently?"

"If so, do you really think the Bean Counter would have told me that? I am the police commissioner, if you recall."

"I do remember hearing something about that."

"You'd do well not to forget," Dino said.

"I wonder how long the 'foreseeable future' is."

"At least long enough for us to have dinner tonight. Clarke's at seven?"

"That all depends."

"On what?"

"If I get out of bed in time."

Stone hung up, set the phone down, and stretched out next to Carly, who soon made it clear his odds of getting out of bed for dinner were low.

Keep reading for an exciting excerpt from

OBSESSION

a Teddy Fay novel featuring Stone Barrington,
by Stuart Woods and Brett Battles.

CHAPTER 1

Teddy Fay flicked his gaze to the rearview mirror and grimaced. The gray sedan was still following him.

Returning his attention to the blacktop ahead, he searched for a turnoff to another road he could use to help him lose his tail. But there wasn't even a wide enough spot where he could pull over. For as far as he could see, the road in front of him was squeezed between the rise of a mountain slope on one side and its descent on the other.

When he glanced again at the mirror, he caught sight of the fear in his own eyes. Fear that this lonely mountain road would be where he died.

He squeezed the steering wheel and shook the thought away. He couldn't let them win. Would not let them win.

He glanced again at the car in the mirror. "Go ahead.

Give me your best shot. I will not let you get to me today."

With renewed determination, he pressed down hard on the gas. For the first time since he'd realized he was being followed, the gap between the other vehicle and his grew instead of shrinking.

If he could, he would have gone even faster, but he knew if he did, the force of a turn might send him crashing into the hillside or flying over the drop-off.

He rounded another ridge, hoping this time he'd spot a turnoff he could use to escape, but the twisting road continued as far as he could see.

He swallowed back his fear. "You can do this. You have to. For them."

He leaned toward the steering wheel, his eyes glued to the road ahead just as the other car came into view behind him.

"Cut." Peter Barrington's voice came through a speaker mounted under the dash of Teddy's car. "I think we got it, Mark."

The tension in Teddy's face vanished. "Excellent. If you're happy, I'm happy."

Teddy released the steering wheel and leaned back. His car was being towed behind the camera vehicle that had been filming him, and he'd only been pretending to drive.

Handling the mountain road on his own would have been child's play. But, alas, today he wasn't Teddy Fay, but rather Academy Award–winning actor Mark Weldon, one of his three main identities. The other was that of film producer Billy Barnett.

Thanks to a talent for altering his appearance, honed during years spent in the CIA, only a select few knew that the three men were one and the same.

This trip up the mountain was Teddy's and the crew's fifth of the day. On the first three, a cameraman had been in the seat beside Teddy, filming close-ups and shots of the pursuing sedan. During the last two trips, Peter's crew had been focused on capturing exterior shots. Peter was the film's director, and the only director for whom actor Mark Weldon worked.

The film was called *Storm's Eye*. Its plot centered on a man named Tyler Storm, who'd spent most of his life on the wrong side of the law but was now trying to make amends for his past deeds. Peter had written the part of Storm with Mark in mind, a lead role to celebrate Mark's status as a major award winner.

The camera truck towed Teddy's car into a dirt parking lot where several other production vehicles waited, as well as Peter, who was standing behind the television on which he'd been watching the camera feed.

When Teddy climbed out of the car, he gave Luke Reed, the man in charge of the rig carrying Teddy's sedan, a nod. "Thanks for not driving us off the cliff."

"They have to pay extra for that."

"Then I'm thankful we have a cheap producer."

"Ha, I'll tell Billy you said that when I see him."

"Trust me, it's not anything he doesn't already know."

Teddy smiled to himself. If only Luke knew. When Teddy reached Peter, who was reviewing what they had shot, he glanced at the monitor. "Everything look good?"

"See for yourself."

Peter pressed a button and the footage from the last trip up the mountain began to play. Teddy had no ego when it came to his acting. All he wanted to do was a good job. When he watched himself, he did so rationally, looking for how he could improve his performance.

The direct shots of Teddy were full of tension and worry, while the footage of the pursuing sedan had the right sense of urgency he knew Peter was after. As he watched, he picked out a few moments in which he felt he could have done a better job, but he was pleased overall.

Peter smiled as the playback ended. "I'd say you're starting to get the hang of this acting thing."

"Maybe someday I can make a living at it."

"It's always good to have goals, even for you."

With his lead role in *Storm's Eye*, Teddy was making enough from Mark Weldon's acting work to live quite well. Given the even larger amount he earned as Billy Barnett, and the tidy sum he'd already amassed before entering the film business, Teddy was more than comfortable.

"Do we need to go again?" he asked.

Peter clapped him on the back. "I've got what I need. You're wrapped for today. Go back to the hotel. Grab a drink and enjoy some time by the pool."

"I like the way you think."

"Then you'll love this. If I'm not mistaken, you're off tomorrow morning, too."

"You're full of good news today."

Peter chuckled in a not-so-fast kind of way. "I was thinking that if you see Billy, you could tell him tomorrow morning might be a good time for a set visit." Unlike Luke the camera truck driver, Peter knew all about Teddy's identities.

"If I happen to run into him, I'll let him know."

CHAPTER 2

Teddy felt the familiar sensation of being watched the moment he stepped into the lobby of the Santa Barbara Hills Hotel.

At least two-thirds of the patrons were sneaking glances at him. Since receiving his Oscar for best supporting actor, Mark Weldon could no longer roam the streets—or in this case, walk through a hotel lobby—unrecognized. It was all rather unfortunate. The persona had served Teddy well when no one had known who Mark was. But gone were the days of conducting any truly covert work as the actor. He had no choice now but to limit Mark's appearances to shoot days, like today, and big events at which Mark's attendance was expected.

"Hello, Mark."

Teddy turned toward the familiar voice of Ben Bacchetti, head of Centurion Studios, the studio behind all of Peter

Barrington's movies. Like their fathers, Peter and Ben were best friends. Some friendships might be strained by working together, but theirs had only grown stronger.

Ben walked toward Teddy from the direction of the hotel restaurant, in the company of Peter's father, Stone Barrington. A successful lawyer with a reputation for being an expert at handling difficult situations, Stone was on Centurion's board of directors. He was also the man responsible for Teddy working with Ben and Peter.

"Ben, Stone, isn't this a pleasant surprise."

"Good to see you, Mark," Stone said.

"Here to check up on the shoot?"

Ben shook his head. "Not this trip. We're here for a meeting with a production company looking for a studio partner. I'll be heading back to L.A. when we're done."

"Don't production companies typically come to you?"

"This isn't a typical production company. Have you heard of Carl Novak?"

"The tech billionaire?"

"That's the one. You've probably seen his wife before, too. Rebecca Novak? She used to be a model. You name a fashion magazine, and she's been on the cover."

"They want to get into the film business?"

"They already have a few productions underway, documentaries and films that deal with problems that don't get a lot of attention. Without a studio partner like us, they'd be stuck making only a film or two a year. They'd like to do more than that. We're one of several groups they're talking to." Ben's phone buzzed. He checked the screen. "Our car's here. We'll see you later, Mark."

Ben headed for the exit, but Stone turned to Teddy instead of immediately following. "Could you do me a favor?"

Teddy arched an eyebrow. "Depends on the favor."

"If you run into Billy Barnett, can you tell him I'd love to have dinner with him? Say, seven-thirty?"

Teddy twisted his head to the side, attempting to look at his back.

"What are you doing?"

"I'm trying to see if someone put a Billy Barnett's personal assistant sign on my back. He has a secretary and an *actual* personal assistant now, you know. You could call one of them, and they'd let him know."

Stone smiled and patted him on the arm. "I figure this way's faster."

CHAPTER 3

Across town, the owner of Smiling Eyes Flower Shop tore off the receipt and held it out. "Thank you, Mrs. Novak. I'll make sure they're delivered to your house by one p.m. tomorrow."

"Perfect." Rebecca Novak glanced at the clock on the wall. It was already a quarter to three. "Is that the right time?"

The clerk looked to see what she was talking about and nodded. "Within a minute or two, I think."

"I didn't realize it was so late. Thank you again for your help." Rebecca returned her wallet to her purse and hurried out the door.

She'd spent more time in the shop than she'd planned to and would need to rush to get home in time for the start of the meeting with Centurion Studios.

As she approached her car, she noticed the passenger

door of the sedan beside hers was open, blocking her way. A man was leaning into the other vehicle, as if he was getting something out of the glove compartment.

At any other time, she would have patiently waited for him to finish, but she was already running behind. "Excuse me."

The man didn't seem to hear her.

"Sir? If you don't mind, I need to get into my car."

He looked over his shoulder and then at her car, as if only now realizing he was in her way. "Oh, I am so sorry."

He closed his door and stepped to the side to clear the path.

Once he was out of the way, she flashed him a smile and moved past him. "Thank you."

As she reached for the handle of her door, she sensed a sudden presence behind her, then felt something hard jab into her ribs. Startled, she looked down and saw the muzzle of a gun pressing against her.

"If you do anything to call attention to us, I will kill you."

She stared at him, wide-eyed, then started to pull her purse from her shoulder. "Whatever cash I have, it's yours."

He shoved the gun harder against her side. "You only speak when I ask a question. Otherwise, you remain silent. Nod if you understand."

"What?"

He ground his gun into her again. "Understand?"

She squeezed her lips together and nodded.

"Good. Take a step back and open the door to my car."

Her eyes widened. But before he could twist the gun into her again, she did as he'd asked.

"Now, get in."

When she didn't move right away, he said, "I can either kill you here or take you with me. Either way the message will be sent. The choice is yours."

She hesitated another second before ducking down and climbing into the seat. The instant she was inside, someone who'd been hiding in the back grabbed her arms. She tried to struggle free, but a wet cloth flopped over her mouth and nose. Rebecca twisted her head back and forth to get out from under it, but a hand held it in place.

Her head began spinning as a wave of vertigo washed over her. She tried to fight it, but with every breath she took she felt her consciousness slipping further away.

Though she heard her captors speaking, their voices were muted by the thick cloud of confusion churning in her head, and it took her a second to realize they weren't speaking in English. Yet she still found she understood a word or two here and there.

"Get her key," one of the voices said.

She felt her purse being tugged from her shoulder, but she could do nothing about it. Around her, the world grew dark as her vision narrowed.

"That's right, Mrs. Novak. Go to sleep."

As the last of her consciousness began to fade away, she realized why the language sounded so familiar.

They were speaking in her husband's native Croatian.

Zoran Janic tried very hard not to glance at his phone. His cousin, Neno Bilic, was supposed to have contacted him an hour ago. If something had gone wrong with the operation, Janic would not be pleased.

Two weeks ago, he had sent Neno, along with two of Janic's other men, Sava Kordo and Pavel Dodic, to America to deal with a problem Janic had neglected for too long. Fourteen days should have been more than enough time to rattle the Novaks in preparation for the main event. Which, if things had gone well, should have begun within the last couple of hours.

Janic walked to his window, his jaw tense. Those nosy documentary filmmakers were irritating enough. After they had shown up in Croatia and started digging up information on crime networks in the Balkans, networks that with just enough poking would lead right to him, he found an easy enough way to shut down production. But his plans changed to something bigger when he learned Carl Novak and his wife, Rebecca, were the ones backing the film.

It was a sign that the time for payback had arrived, a sign Janic had been waiting years for.

Finally, his phone rang.

"Yes?"

"It's me," Neno said. "We have her."

All the tension Janic had been holding on to vanished. "Any problems?"

"None."

"No one saw you?"

"No one."

"Good work. Proceed as planned."

Janic ended the call, grabbed the bottle of whiskey off the shelf behind him, and poured himself a celebratory drink.

STUART WOODS

"Addictive . . . Pick it up at your peril. You can get hooked."

—Lincoln Journal Star